Space Debris Peril: Pathways to Opportunities

Space Debris Peril: Pathways to Opportunities

Capacity Building in the New Space Era

M. Madi

O. Sokolova

CRC Press

Taylor & Francis Group

Boca Raton London New York

CRC Press is an imprint of the
Taylor & Francis Group, an **informa** business

First edition published 2021
by CRC Press
6000 Broken Sound Parkway NW, Suite 300, Boca Raton, FL 33487-2742

and by CRC Press
2 Park Square, Milton Park, Abingdon, Oxon, OX14 4RN

Library of Congress Cataloging-in-Publication Data

Names: Madi, Matteo, editor.
Title: Space debris peril : pathways to opportunities : capacity building
in the new space era / [edited by] M. Madi.
Description: Boca Raton, FL : CRC press, 2020. | Includes bibliographical
references and index.
Identifiers: LCCN 2020028363 | ISBN 9780367469450 (hardback) | ISBN
9781003033899 (ebook)
Subjects: LCSH: Space debris.
Classification: LCC TL1499 .S65 2020 | DDC 363.72/80919--dc23
LC record available at https://lccn.loc.gov/2020028363

ISBN: 9780367469450 (hbk)
ISBN: 9781003033899 (ebk)

Typeset in Computer Modern font
by Cenveo Publisher Services

Visit the eResources: www.routledge.com/9780367469450

To be mindful.

Contents

SECTION III Dealing with Space Debris. Sociotechnical Concerns

CHAPTER   3 ▪ Space Debris Sustainability: Understanding and Engaging Outer Space Environments   69

MICHAEL CLORMANN and NINA KLIMBURG-WITJES

SECTION IV Technological Challenges & Current Developments

CHAPTER   4 ▪ Overview of the Proposals for Space Debris De/Re-Orbiting from the Most Populated Orbits   85

ANDREY A. BARANOV and DMITRIY A. GRISHKO

SHINICHI KIMURA

SECTION V   **Regulation & Legislation**

CHAPTER   6 ▪ Addressing the Inevitable: Legal and Policy Issues Related to Space Debris Mitigation and Remediation   137

LUCY STEWARDSON and STEVEN FREELAND

SECTION VI **Risk Complexity and Space Debris Governance**

OLGA SOKOLOVA and MATTEO MADI

SECTION VII **Conclusion**

MATTEO MADI

List of Figures

List of Tables

Foreword

Space is infinite. We can place satellites, spacecraft, and manned space stations in space as much as we like.... Is this belief, which we have been considering to be true for a long time, still correct?

The recent space debris threat is leading many to question this belief. Space, especially the space near Earth is not infinite anymore.

Space has been providing us with a "resource of room" where we can place satellites and other artefacts. Just like the resources on the Earth such as petroleum, coal, uranium, or rare metal, this "resource of room" of space is not unlimited. If we continually spend the resource without care, it will be expired eventually, and the expiration date may not be so far away.

Some of you may witness the debris peril in the movie "Gravity." The debris generated by the collision of communication satellites damaged the Space Shuttle and jeopardized the crews, and the debris generated in the same event also destroyed the International Space Station. This kind of collision, which happened in the real world in 2009, created more than 1,000 articles of debris. The speed of flying objects in space, especially the relative speeds of two flying objects sometimes exceeds 10 km/s, which transforms even cm sized miniature fragments into dangerous bullets. Besides, though large debris with more than 10 cm size can be monitored from the ground, smaller debris cannot be seen, which can be very dangerous, an "unseen threat."

However, most of us do not know how severe the debris threat is now, and how this threat will enlarge in future. If the space surrounding the Earth becomes full of debris, not only will satellites be unable to be safely placed in orbits, but also human space activities will be very much impeded. We believe that in the future human beings will travel and live in space, but the debris around the Earth will be a huge hazard to such human advances into space.

It may be true that the recent rapid increase of the number of small satellites accelerates the debris threat, but we should also keep in mind that the objects launched into the space in the past, including rocket upper stages, satellites of small to large size, and jettisoned objects from various activities such as adapters between satellites and rockets can be sources of debris; that is, small satellites are not only the villains. We should get correct knowledge about debris; what the level of debris threat is right now, which type of debris will be the big threat to general space activities, what will be the most appropriate actions to control them.

We engineers who are working on satellites should observe the rule that satellites have to be removed from orbits after their missions are completed. However, sometimes this countermeasure may not be enough, and in such cases we should "sweep the space." This is the concept of Active Debris Removal, which imposes upon us many technological challenges such as how to change orbit to make a rendezvous with the target debris, how to capture the usually tumbling objects, and after the capture how to make the debris descend to the atmosphere to be burnt out. These problems will eventually be solved technologically, but the difficult issue is who will willingly pay for the cost of removal of the old satellites or rockets which still remain in orbits after the end of their life time. International discussion, negotiations and proposals to solve legal issues would be needed to effectively find solutions to these delicate issues.

The most important things for us to do now are to know the true situation of space debris and what kinds of options exist to solve the issues. I believe that this book will help you grasp the whole picture of the current space debris issue. Space situation awareness and space traffic management, the two important concepts now pursued in the international framework, will be explained and discussed in detail. Two chapters are dedicated to engineering aspects of space debris issue, that is, how to remove the debris from the orbits. The next part will focus on legal and policy issues related to space debris, including the international guidelines created by Inter-agency community and United Nations, followed by the assessment of space debris and review of its remediation projects. It is also discussed how space debris issue will impede human outer space activities in the coming new space age and sketch out potential opportunities of societal engagement into this problem.

This book provides up-to-date knowledge of space debris and valuable insights on how to grapple with this issue from legal, technical, economical and societal aspects. I would strongly recommend that everyone who is working on space development and utilizations and even non-specialists once read this book and think over how human being should be faced with this issue.

Tokyo, Japan, May 2020   *Shinichi Nakasuka, Ph.D., Professor*
                                          University of Tokyo

Preface

The change in space activities in the New Space era has turned a new page in the space-borne risks profile. The space debris threat is considered as an emerging risk threatening our modern society's security and well-being. Thereby, increasing social awareness on this topic is mandatory. Following this concern, this book project is launched in order to cover the whole spectrum of the space debris problem, from legal to socio-economical aspects, and also to satisfy the need for a solid reference on current trends in legislation preparation and technological development for the space debris problem across the world, including the USA, Japan, Russia, Australia and European countries.

This book shows how space debris risks evolved over time and outlines the current challenges to the community from legal, political, economic, social and technical realms. This manuscript represents an attempt to describe what the credible risk is and to give insights into how to mitigate the negative consequences. The authors are affiliated with world-renowned universities, research institutes, private commercial sectors active in the field of space debris mitigation, space policy and law, and space situational awareness. The target audience is among the general public, scientific communities, policy makers, business developers, (re)insurers and international standards developers for space operations and orbital debris mitigation. Furthermore, since the topics of space debris and the importance of Clean Space are evolving and gradually becoming of interest to the public, it is expected that the intended book will attract a broader audience among non-specialists in various sectors.

The book starts with describing the nature of space debris as a peril; a historical overview of the problem and the current vision. Dan Oltrogge and James Cooper, respectively from the Center for Space Standards and Innovation and the Commercial Space Operations Center at Analytical Graphics Inc. (AGI) give an insight into the concepts of Space Situational Awareness and Space Traffic Management (STM). This knowledge is supported by discussion on the sociotechnical perspective of dealing with space debris given by Michael Clormann from Technical University of Munich and Nina Klimburg-Witjes from University of Vienna. This takes the reader to the section on Technological Challenges and Current Developments, which consists of two chapters. First the technical chapter authored by Russian senior researchers, Andrey Baranov and Dmitriy Grishko, gives an overview of the technical proposals for space debris removal from various orbits; thereafter Shinichi Kimura, Professor at the Tokyo University of Science, enlightens the reader as to the state-of-the-art space debris mitigation methods based on Commercial-Off-The-Shelf (COTS) technologies. The follow-up section is focused on legal and policy issues. Steven Freeland from the School of Law at Western Sydney University and Lucy Stewardson from The Brussels Bar address the inevitable legal and policy issues related to space debris mitigation and remediation. The senior lawyer from Paris, Cécile Gaubert, summarizes the risk chain created by the space debris problem and solutions for different stakeholders, notably the insurance market, to remediate such risks. The risk analyst, Olga Sokolova, speaks on the problem of space sector resilience assessment and the governance of risks associated with the space debris. At last, it is shown how the space debris peril creates new opportunities to enrich our modern society's resilience from legal, technical, economical and societal standpoints.

The hardcopy book readers can access the colourful versions of the greyscale figures contained in this manuscript on the book product webpage, accessible on: `https://www.routledge.com/9780367469450`. Besides, the eBook contains the colourful figures.

Zürich, Switzerland, May 2020
Matteo Madi, Dr. sc. ETH
Sirin Orbital Systems AG

Contributors

Andrey A. Baranov
Keldysh Institute of Applied
    Mathematics of RAS
Moscow, Russia

Michael Clormann
Munich Center for Technology in
    Society,
    Technical University of Munich
Munich, Germany

Jim Cooper
Commercial Space Operations Center,
    Analytical Graphics, Inc. (AGI)
Exton, Pennsylvania USA

Steven Freeland
School Of Law, Western Sydney
    University School of Law
Sydney, Australia

Cécile Gaubert
Gaubert Law Firm
Paris, France

Dmitriy A. Grishko
Bauman Moscow State Technical
    University (BMSTU)
Moscow, Russia

Shinichi Kimura
Space Robotics & Small Satellite
    Technologies, Tokyo University of
    Science
Tokyo, Japan

Nina Klimburg-Witjes
Department of Science and Technology
    Studies, Vienna University
Vienna, Austria

Matteo Madi
Sirin Orbital Systems AG
Zürich, Switzerland

Dan Oltrogge
Center for Space Standards and
    Innovation,
    Analytical Graphics, Inc. (AGI)
Colorado Springs, Colorado USA

Olga Sokolova
Sirin Orbital Systems AG
Zürich, Switzerland

Lucy Stewardson
The Brussels Bar
Brussels, Belgium

About the Contributors

Andrey A. Baranov is a leading researcher of Keldysh Institute of Applied Mathematics, Russian Academy of Sciences, Space Flight Dynamics Department, Dr.Sc. (Phys-Math) and Professor of Peoples' Friendship University of Russia. His research interests lie in the area of orbital manoeuvrers optimization and orbit maintenance. He also focuses on on-orbit servicing, active space debris removal missions and satellite collision avoidance, near-Earth and deep space missions design. He developed a new theory of how to optimize parameters of orbital manoeuvrers that is characterized by wide application and international recognition. For example, the results of this new theory were used in ATV missions. Andrey Baranov is a strong contributor to space exploration. He has been participating in flight control of more than 140 different spacecraft. Andrey is the IAA corresponding member, and has received the Badge of Honour order (Russia) and Medal to 2nd degree Order of Merit (Russia). He is an author of more than 100 publications, 2 books and 2 patents.

Michael Clormann, M.A., is a scientific associate and doctoral candidate at the Munich Center for Technology in Society, Technical University of Munich. His research is focused on the sociotechnical implications of currently progressing transformations under the umbrella term "New Space Age". He specializes in considering the societal implications of sustainability concerning the rising challenge of space debris. Additionally, he explores the changes in institutional and political innovation culture within the European space sector. He is co-founder of the Social Studies of Outer Space research network.

Jim Cooper joined AGI in August of 2016 as a Senior Systems Engineer supporting the ComSpOC/Space Situational Awareness Solutions Business Unit for US and international government organizations. He supports strategy and business execution to deliver commercial SSA solutions to government and military organizations worldwide, including the strategic pursuit and development of large, long-term program opportunities and enterprise accounts. Mr. Cooper has over 30 years of professional experience in SSA policy, operations, international engagement, and funding support. Prior to joining AGI, Mr. Cooper provided SETA support for 16 years to HQ USAF/A3 and for 3 years to HQ AFSPC/A3 in the SSA mission area. In this capacity, he advocated for SSA in the Planning, Programming, Budgeting and Execution (PPBE) process, supported SSA operational concerns and policy development, and conducted international engagement in SSA for the AF/A3. After graduating with a B.S. from the US Air Force Academy in 1985, Mr. Cooper served in the USAF for 8 years as an Orbital Analyst in the Space Surveillance Center at Cheyenne Mountain AFB, Colorado and a Master Instructor for Undergraduate Space Training at Lowry AFB, Colorado. Upon separating from the USAF, he worked as a commercial diver for 4 years in oilfield support and inland marine facility inspection. Mr. Cooper is married to Mrs. Sarah Cooper; they are the parents of two children – Zachary, 13 years old, and Elizabeth, 9 years old.

Steven Freeland is Professor of International Law at Western Sydney University, Australia. He is also Visiting Professor at the University of Vienna, Permanent Visiting Professor at the iCourts Centre of Excellence for International Courts, University of Copenhagen, Adjunct Professor at the University of Hong Kong, Senior Fellow at the London Institute of Space Policy and Law, Visiting Professor at Université Toulouse1 Capitole, and Associate Member at the Centre for Research in Air and Space Law, McGill University. He has represented the Australian Government at United Nations Conferences on space, and has advised various governments on issues related to the national and international regulation of space activities and the development of a national space-industry strategy. He has been appointed by the United Nations Committee on the Peaceful Uses of Outer Space to co-lead multilateral discussions regarding the exploration, exploitation and utilization of space resources, and by the Australian Government as a Member of the Australian Space Agency Advisory Group. Among other appointments, he is a Director of the International Institute of Space Law, and a member of the Space Law Committees of both the International Law Association and the International Bar Association.

Cécile Gaubert is a lawyer registered at the Paris Bar – France. She is acting within the fields of aviation and space activities. Cécile is a former Head of Legal and Claims – Aviation & Space Department of Marsh France. Cécile holds a Masters degree of Law from the University of Paris XI and a Certificate from the European Centre for Space Law's Summer Course. She is currently chairwoman of the space committee of the French association on air and space law (Association Française de Droit Aérien et Spatial). Cécile has published several articles relating to space insurance and has given speeches and presentations to colloquia on specific topics such as small satellites, GNSS signal liability, space tourism, space debris and others. Cécile is a member of different space related institutes including IDEST (Institut pour le Développement du Droit de l'Espace et des telecommunications) and IISL (International Institute for Space Law). She is also invited to French universities for lectures on space and aviation insurance.

Dmitriy A. Grishko is an Associate Professor of Bauman Moscow State Technical University, Department of Theoretical Mechanics, PhD (Phys-Math). His research interests lie in the area of orbital maneuvers optimization and space flight dynamics. He also focuses on active space debris removal missions and satellite constellations. He developed a methodology, that permits the composition of effective flyby schemes in the framework of active debris removal missions in LEO and GEO. Dmitriy is honoured with the S.P. Korolev medal (Russia) and "Best Young Professor" Badge of Bauman Moscow State Technical University. He is an author of more than 25 publications and 1 patent.

Shinichi Kimura received the B. S. degree in pharmacology from University of Tokyo, Tokyo, Japan, in 1988 and M. S. and Ph. D. degrees in pharmacology from the graduate school of University of Tokyo, Tokyo, Japan, in 1990 and 1993 respectively. In 1993 he joined the Communications Research Laboratory (which was converted to the National Institute of Information and Communication Technology in 2004) and has been the group leader of Smart Satellite Technology Group since 2004. In 2007, he moved to Tokyo University of Science as an associate professor, and in 2012 he became professor. He has been engaged in space robotics and autonomous control technologies to maintain telecommunication satellites. He made experiments on Manipulator Flight Demonstration on STS-87, Engineering Test Satellite VII which is the first tele-robotic satellite, visual analysis experiment of deployable antenna (LDREX), Micro-LabSat, and REXJ (Robot Experiment on JEM). He developed various on-board equipment such as the visual monitoring system for the IKAROS (Interplanetary Kite-craft Accelerated by Radiation Of the Sun) and asteroid explorer Hayabusa-2. He is engaged in various research and development involving technological issues on space debris mitigation. His technologies have also been utilized in space debris mitigation missions of private sectors such as Astroscale and KHI.

Nina Klimburg-Witjes is a university assistant and postdoctoral researcher at the Department of Science and Technology Studies (STS), University of Vienna. In her work at the intersection of STS and Critical Security Studies, she explores the role of technological innovation and knowledge practices in securitization processes. Tracing the entanglements between industries, political institutions, and users, Nina is interested in how visions about sociotechnical vulnerabilities are co-produced with security devices and policy, and how novel security technologies interact with issues of privacy and democracy. Nina has published extensively on satellite imagery in the context of security issues, has recently edited a book on Sensors as Transnational Security Infrastructures (forthcoming 2020) and is currently working on a monograph on the European launcher Ariane.

Matteo Madi is an entrepreneur, business developer and space-tech specialist with over ten years of work experiences in the Swiss and International public and private sectors. He participated in development processes for various space missions as lead R&D engineer or project manager. Matteo Madi received his M.Sc. (2011) in Electrical Engineering and Space Technologies and his Ph.D. (2016) from the Swiss Federal Institute of Technology Lausanne (EPFL). Mr. Madi worked at the European Space Research and Technology Centre (ESTEC) in 2016 in the frame of ESA's Networking/Partnering Initiative (NPI) program focusing his research on the key technologies enabling compact imaging spectrometers. He is designated as inventor by the European Patent Office for three granted patents, one of which is a joint patent supported by the Technology Transfer & Business Incubation Office (TTBO) at ESTEC, ESA. As business developer, he has been active in providing consultation, strategic advice, and market development support to venture capitals, start-ups, Swiss and international institutions, and contributed to their efforts to align policies with the latest market/technology trends. In response to the new demands of the emerging space market, in 2019, Matteo Madi founded Sirin Orbital Systems AG based in Zürich, Switzerland, aiming at developing enabling technologies for On-Orbit Servicing (OOS) of satellites.

Dan Oltrogge is the Director of the Center for Space Standards and Innovation at Analytical Graphics Inc. (AGI). Mr. Oltrogge led the formation of and serves as administrator for the Space Safety Coalition. He is also the program manager of the Space Data Center, a lead policy and analysis expert for AGI's Commercial Space Operations Center (ComSpOC), and a frequent author of technical papers and peer-reviewed journal articles. Mr. Oltrogge has contributed to the development of numerous space operations and debris mitigation international standards and best practices under the auspices of

ISO, CCSDS, CONFERS, ANSI, IAA, and industry associations. His technical focus areas include space debris, launch and orbital operations, collision avoidance, RF interference mitigation, Space Situational Awareness and Space Traffic Coordination and Management.

Olga Sokolova is a Risk Analyst with Engineering and Business Administration background. She received her Ph.D. degree in 2017 from St. Petersburg Polytechnic University. Her research interests lie in the field of critical infrastructure risk assessment for natural and technical hazards. She has been engaged in development and analysis of structural risk-management tools towards a sustainable future which can be used by the general public. Olga has cooperated with various entities including Swiss Reinsurance Company Ltd. (Swiss Re), Swiss Federal Institute of Technology Lausanne (EPFL), Paul Scherrer Institut (PSI), and Stanford University. In 2014, she published a lead-authored brochure published by Swiss Re on space weather challenges and threats to power grid operators and industry. She has a record in raising social awareness of spaceborne risks and opportunities brought to society by the "New Space" industry development. Olga regularly shares her thoughts on space industry risks at corresponding events. Ms. Sokolova contributes in developing ad-hoc risk models validations, challenging the appropriateness of assumptions and modelling processes, and risk governance.

Lucy Stewardson obtained her Master of Law from the Université Libre de Bruxelles, Belgium, where she specialised in Public and International Law. She wrote her final year paper on the international regulation of space debris and conducted further research in the field of space law after graduation. Lucy is currently working as a lawyer in international arbitration at The Brussels Bar.

I

Introduction

CHAPTER 1

Space Debris, a Peril

Olga Sokolova

P ERIL is literally defined as a serious and immediate danger. The public awareness of space hazards was historically the concern of a limited number of countries, who had operational scientific space missions. However, the increase of activities in the New Space era and the growing volume and variety of ground-based infrastructures dependent on space systems change the picture. The New Space infrastructure is evolving as a new backbone system, the reliable operation of which is decisive for society's well-being and economic stability. Among spaceborne risks, space weather is the one which has the highest awareness among the general public. Back in 2011, the Organisation for Economic Co-operation and Development (OECD) defined space weather as a future global shock. As an outcome, a group of countries established programs for assessing the risk and included it in their national risk portfolio. The aforementioned group of countries includes territories together with those traditionally considered as "high-risk" zones and the countries with "low-risk". Public discussions supported by scientific studies led to changes in the understanding of risk and its consequences.

The space debris risk is considered as an emerging risk. This is based on the following promises: the rapid development of the space debris environment, the growing inter-connectivity between spaceborne and ground-based infrastructures, the development of new technologies for problem assessment and mitigation. The quantification of emerging risks is a difficult task, and their potential impacts on businesses are not sufficiently known. In other words, emerging risks can be classified as *a peril that keeps risk managers up at night*. Though the terms "risk" and "peril" are often used interchangeably, the following definitions are considered in this book: "risk" is an uncertainty leading to adverse consequences; "peril" is a potential cause of a loss or damage; and "hazard" is a danger from a peril.

In recent years, the space industry has undergone profound changes. The new actors in the space sector work tirelessly to effect a transformation in the space industry similar to that which occurred previously in the aviation industry. At the same time, the standardization and industrialization of the production of small-satellites enable new innovative business models and missions. The space industry has evolved towards the predominantly commercial operation. In the New Space era, it is essential to prepare not only the active players but also the public for the challenges in front of the future space activities which are taking place with no political or geographical frontiers.

It is a sobering fact that the "zero-risk" level does not exist. Peril assessment is important for designing mitigation solutions, since the evaluation of risk provides a basis for planning and allocation of limited resources: technical, financial and others. However, it is not the first time in history that society has faced the challenge of finding solutions to assess an emerging risk. For instance, two major risk analysis tools, namely HAZOP and FMEA, were originally developed in support of nuclear industry development, when it was in its infancy during the 1960's. Above and beyond giving fundamental information on the space debris problem from various legal, technical and economical perspectives, this book presents an attempt to describe our knowledge on the space debris problem based on a realistic worst-case scenario, for which modern society should prepare.

Experience shows that the time after a disastrous event is the window of opportunity for implementing actions. The relevant stakeholders – in order to avoid the same loss in the future – are eager to implement actions with higher cost in order to boost resilience. However, experiences of recent catastrophic event mitigation such as the eruption of Eyjafjallajökull volcano in 2010, the Japan tsunami in 2011, hurricane Sandy in 2012 proved that prevention is more beneficial than mitigation. Therefore, this moment in time can be considered as a window of opportunity for developing common approaches for dealing with space debris peril. The range of technical solutions supported by economical, societal and legal opportunities are the focus of this book project.

This manuscript has a strong interdisciplinary emphasis, focusing on issues related to space debris risk determination, evaluation, and management. Each chapter is meant to be understandable relatively independent of the others. The chapters are written by leading experts from many fields dealing with the space debris problem. The contents of the book can be briefly described as follows:

CHAPTER 2 – by D. Oltrogge and J. Cooper describes the "virtuous cycle" nature of the wholly inter-connected processes of space situational awareness and space traffic management. This chapter challenges the legal, financial and international engagement aspects of space situational awareness in regards to our ability for collision mitigation.

CHAPTER 3 – by M. Clormann and N. Klimburg-Witjes shows how social awareness of space debris peril has been raised through media coverage, outreach activities and relevant stakeholder concerns. It describes how space debris risk is understood as a socio-technological challenge and which instruments are used for building societal resilience. The authors hope that their contribution will provide invaluable insights for scientists, engineers, risk analysts, policy-makers, and the general public on the impact of space debris on modern society.

CHAPTER 4 – by A.A. Baranov and D.A. Grishko gives an overview of space debris mitigation approaches. It also provides a survey of both planned and implemented projects aimed at demonstrating the possibility of removing objects to disposal orbits or spacecraft repairing.

CHAPTER 5 – by S. Kimura concerns the problem of active debris removal technology development. It represents the real example of how a low-cost intelligent guidance and navigation system can be developed for enabling reliable space debris mitigation.

CHAPTER 6 – by L. Stewardson and S. Freeland reviews the current status of legal and policy issues for space debris mitigation. Debris mitigation and, in particular, remediation measures raise many technical, economic and political controversies. This chapter describes the history of the legal framework related to space activities, raises questions that require careful consideration and shows opportunities related to space debris regulation.

CHAPTER 7 – by C. Gaubert discusses space debris as an issue in terms of legal risks. It shows which instruments can be used and which opportunities exist to transfer this risk to dedicated markets (such as insurance) or States. The uniqueness of the chapter is driven by the fact that it not only focuses on risks created by space debris, but also on risks associated with remediation projects that may create new risks.

CHAPTER 8 – by O. Sokolova and M. Madi stresses the problem of space system resilience assessment and questions relevant stakeholders on adopting methodologies for its assessment and how to measure "success". The recommendations for enhancing a sector's resilience and governance options are given. It is also shown how the current legal framework pushes forward the strategic thinking of the resilience concept in the New Space era.

This book will not specifically address the basic definition of the nature of space debris, how they are created and evolved. The readers are highly recommended to refer to the following resources authored by experts in the field to acquire background information: Chapter 4, written by T. Schildknecht, in Nicollier, C., & Gass, V. (2016). *Our Space Environment, Opportunities, Stakes and Dangers.* CRC Press; Pelton, J. N. (2013). *Space Debris and Other Threats from Outer Space.* Springer; and The European Space Agency (ESA)'s Clean Space page entitled, *About Space Debris*, accessible on: `https://www.esa.int/Safety_Security/Space_Debris/About_space_debris`.

II

Shaping Knowledge about Our Space Operating Environment

CHAPTER 2

Space Situational Awareness & Space Traffic Management

Dan Oltrogge

Center for Space Standards and Innovation, Analytical Graphics Inc., Colorado Springs, Colorado USA

James Cooper

Commercial Space Operations Center, Analytical Graphics Inc., Exton, Pennsylvania USA

CONTENTS

"The price of light is less than the cost of darkness."

- Arthur C. Nielsen

THE process of obtaining timely, accurate, comprehensive, and transparent awareness of our space operating environment is known as Space Situational Awareness (SSA). Collision and Radio Frequency Interference (RFI) risks are significant and continue to increase, particularly in the New Space era of large constellations, increasingly smaller spacecraft and cheap access to space. In this chapter, we examine what makes SSA challenging from the standpoint of policy, finance, and international engagement, and how such challenges impede our ability to mitigate collision and RFI risks. This is necessary for the long-term sustainability of the space environment. SSA is also a key enabler for space governance, whether it be spacecraft operator self-regulation, treaties and guidelines of the United Nations and the Inter-agency Space Debris Coordination Committee (IADC), international standards of the International Organization of Standardization (ISO) and the Consultative Committee for Space Data Systems (CCSDS), or national policy and regulatory governance. The "virtuous cycle" nature of these wholly inter-connected governance instruments is discussed. The topic of Space Traffic Management (STM) is also examined. We examine why timely and substantive improvements are required across the SSA and STM enterprises to maximize the effectiveness of safety-of-flight procedures and analyses. Advances in commercial and government SSA analysis tools, algorithms, and tracking sensors are discussed that help facilitate these required enhancements.

2.1   INTRODUCTION TO SSA AND STM

Space Situational Awareness and Space Traffic Management are foundational to enabling spaceflight safety, space sustainability, and security. In this chapter, we define and explain the basic components of SSA and STM, including space object tracking, algorithms, close approach assessment, and the spacecraft operator decision-making process. We examine how technical challenges and diverse perspectives can affect SSA and STM services, adversely impacting the collision and RFI risk mitigation necessary for a functional space environment.

We then describe the providers of SSA and STM services and the pivotal roles these services play in promoting the long-term sustainability of outer-space activities. We discuss the critical role of advanced algorithms and analytics, data fusion, and international space standards in producing decision-quality SSA, facilitating data exchange, and codifying expected norms of behaviour. Finally, we discuss how the current and anticipated space debris situation indicates that we need to have a sense of urgency and proactively transition to operationally relevant SSA and STM services.

2.2 DEFINING SSA AND STM – WHAT ARE THEY?

SSA and STM are not universally defined across the global space community. Diverse definitions are inevitable given the equivalent diversity of the organizations using them and how those terms relate to these organizations' missions and priorities. Commercial operators, regulators, and national security experts have disparate SSA requirements and priorities. SSA can prevent collisions; provide insight into space and ground capabilities; enable and support national security; and detect, identify, and attribute irresponsible spacecraft behaviours and practices that threaten the long-term sustainability of space activities [1].

2.2.1 Space Situational Awareness (SSA)

At a base level, SSA could simply be defined as an awareness of all relevant activities and objects that can impact or influence one's use of space. SSA could be more extensively defined as "the comprehensive knowledge and understanding of the space and terrestrial environment, factors, and conditions, to include the status of other space objects, radio emissions from ground and/or space transmitters, and terrestrial and space weather, that enables timely, relevant, decision-quality, and accurate assessments, in order to successfully protect space assets and properly execute the function(s) for which a satellite is designed" [2].

The European Space Agency (ESA) defines SSA [3] as "the comprehensive knowledge, understanding, and maintained awareness of: (i) the population of space objects, (ii) the space environment, and (iii) the existing threats and risks."

France, in consultation with its Centre national d'études spatiales (CNES) subject matter experts, considers SSA to be the merger of space weather and Space Surveillance and Tracking (SST) [4].

The European Union's Space Surveillance and Tracking (EUSST) program characterizes SSA [5] as the knowledge of the space environment, including location and function of space objects and space weather phenomena, realized through three pillars: Space Surveillance and Tracking (SST), Space Weather Monitoring and Forecasting, and Near-Earth Objects. The EU defines SST as "the capacity to detect, catalogue, and predict the movements of (man-made) space objects orbiting the Earth."

Others prefer to confine SSA's meaning to past, present, and future positional knowledge of macroscopic objects in near-Earth space. The Space Foundation, an advocate of this limited definition, states [6] that "Space Situational Awareness refers to the ability to view, understand, and predict the physical location of natural and manmade objects in orbit around the Earth, with the objective of avoiding collisions".

These and other SSA definitions are summarized more comprehensively in Tab. 2.1 (not intended as a comprehensive list).

Having acknowledged that a variety of definitions exist, we will adopt for the remainder of this chapter the definition set forth in United States Space Policy Directive 3 (SPD-3) [7]: "Space Situational Awareness shall mean the knowledge and characterization of space objects and their operational environment to support safe, stable, and sustainable space activities."

In recognizing that the military's mission is much more expansive than this safety-oriented definition, the United States Air Force transitioned all of its space organizations over to use of the term "**Space Domain Awareness (SDA)**" in November of 2019. In defence circles, Space Domain Awareness represents not only the catalogue maintenance aspect of some of the narrower SSA definitions, but it also refers to the identification, characterization, and understanding of any factor or behaviour, passive or active, associated with the space domain that could affect space operations, thereby potentially impacting the nation's security, safety, economy, and environment. While Space Domain Awareness is an inclusive term that aligns well with a few of the more comprehensive SSA definitions presented previously, it is primarily relevant to the national security mission and not generally applicable to international initiatives to ensure the Long-Term Sustainability (LTS) of space activities.

2.2.2 Space Traffic Management (STM)

SSA is critical but insufficient to solely support spaceflight safety. To truly meet the needs of space operators and state actors, SSA must facilitate and be complemented by Space Traffic Management (STM) services. An early definition [8,9] of STM from 2006 establishes it as "[...] the set of technical and regulatory provisions for promoting safe access to outer space, operations in outer space, and return from outer space to Earth free from physical or RFI." This definition expressly includes both technical and regulatory aspects, en-

Table 2.1: Comparison of SSA attribute definitions by source

Space Situational Awareness Attribute	AFI 14-SPACE	Alfano, CSSI	France/CNES	ESA	EU	Space Foundation	Space Nav	Space Domain Aware.	US Space Policy	US SPD-3
Characterization of Earth-based space capabilities	●	●						●		
Characterization of space/operating environment	●	●	●				●	●		●
Characterization of space-based capabilities	●	●	●				●	●		●
Comprehensive knowledge and status of space objects	●	●	●	●	●		●	●		●
Current and future knowledge	●	●	●	●	●	●	●	●		●
Identification of bad actors in space								●	●	
Monitoring multinational space readiness	●							●		
Near-Earth Objects (i.e., comets and asteroids)					●			●		
Protects space assets to function as designed		●						●		
Radio emissions (ground- and space-based)	●	●						●		●
Safe, sustainable and stable space activities		●						●		●
Space and terrestrial weather	●	●	●		●			●		●
Space Domain Awareness and analysis	●							●		
Threat monitoring and risk assessment	●		●				●	●		●
Timely, relevant, accurate, actionable		●	●					●		●
Understand & predict space object physical locations	●	●	●	●	●	●	●	●		●

compassing significantly more than just the spaceflight safety analyses that identify a close encounter between space objects (referred to as Conjunction Assessment, or CA). Many prominent Geostationary Earth Orbit (GEO) operators favour this definition; they have significant concerns about RFI and thus seek a definition of STM that couples CA with forensic and predictive RFI analysis capabilities and interfaces.

The breadth of SSA and STM definitions summarized in Tables 2.1 and 2.2 are discussed in more detail in relevant literature [2, 10]. As with SSA, discussion in this chapter will refer to STM as characterized in United States Space Policy Directive 3: "Space Traffic Management shall mean the planning, coordination, and on-orbit synchronization of activities to enhance the safety, stability, and sustainability of operations in the space environment."

While several STM definitions include important regulatory aspects of orbital debris mitigation, none of these definitions encompass the authority and activity of directing a spacecraft to simply "turn left" or "turn right." Rather than directing traffic, a more cogent practical application of STM actions is "coordination" – that is, helping coordinate risks and avoidance manoeuvres between operators. The terms "oversight" and "control" typically denote monitoring compliance and observing, cataloguing, attributing, and monitoring space objects. As a result, "Space Traffic Coordination" (STC) might be a more accurate title for the current collision avoidance process rather than the mainstream phrase, "Space Traffic Management." However, for the sake of this discussion, we shall use the widely-accepted STM term.

In summary, both SSA and STM lack cohesive, universal definitions. Some interpretations of SSA and STM are narrowly limited to the tracking of space objects in order to avoid collisions. The United States stands to benefit from the use of broader, more visionary, and better-balanced definitions that include space weather, RFI, and characterization of capabilities.

The diverse definitions of SSA and STM do have one commonality: they fail to address the lack of coherent traffic management across regimes. The confluence of ground, maritime, and air traffic control (ATC); Unmanned Aircraft System Traffic Management (UTM); high altitude (High E) near-space flight 25,000 feet or more above Mean Sea Level; and the outer space regime is imminent. Launch trajectories already transit or will soon transit all of these regimes. As the burgeoning commercial launch industry launches more spacecraft in the New Space era, all of these regimes will be impacted. It is time to standardize cross-regime traffic management terms that address, prioritize, and integrate safely across all regimes.

Table 2.2: Comparison of STM attribute definitions by source

Space Traffic Management Attribute	Aerospace	Athens Univ.	Blout	DLR	IAA COSMIC	GWU	ITU	NASA/JSC	US Space Policy	US SPD-3
Best practices, standards, tech means	●		●		●					●
Free from physical interference	●		●		●		●			●
Free from RF interference	●		●		●		●			●
Information security								●		
Monitoring and notifications	●		●					●		●
On-orbit collision avoidance	●	●				●		●		●
Plan, coordinate, synchronize activities					●				●	●
Pre-launch risk assessments										●
Safe launch			●		●	●		●	●	●
Safe orbit operations			●		●	●		●	●	●
Safe return from space			●		●	●		●	●	●
SSA			●							
Regulatory/enforcement										
Licensing and allocation					●					
Regulatory			●	●		●				
Rules of the road								●		
Traffic control/enforcement	●	●					●	●		

2.3 WHAT IS SPACE ENVIRONMENT MANAGEMENT (SEM), AND WHERE DOES IT FIT?

Concerned that management of the space environment is not getting the prioritization and focus necessary to ensure long-term sustainability of the space environment, some experts have proposed the term Space Environment Management (SEM) to describe the management of the space debris population. Introducing this term may indeed help the space community and regulators bring focus on this critical yet under-resourced area. While it is clear that more must be done to address the space debris problem, it remains to be seen whether the space community will see the need to make a clear distinction between STM and SEM.

It is our opinion that the current definitions of STM are inclusive of SEM. Consider the SPD-3 definition of STM: *"Space Traffic Management shall mean the planning, coordination, and on-orbit synchronization of activities to enhance the safety, stability, and sustainability of operations in the space environment."*

Note the key phrase *"sustainability of operations in the space environment."* This clearly includes managing the environment (because if one didn't do that, then space operations would no longer be sustainable). From this perspective, we consider that if the term "SEM" is embraced by the space community (as we believe it should), then SEM would become an important component within the overall STM enterprise.

2.4   SHOULD STM INCLUDE REGULATING, MONITORING, AND ENFORCEMENT?

The inclusion of the concept of regulation in some of the above STM definitions is of immediate significance; we are, in fact, far from having any system of harmonized national or multinational regulations [11]:

- Space law [...] currently lacks numerous provisions which are essential for a comprehensive traffic management regime (e.g., pre-launch notification, custody requirements for attribution of collisions, binding regulations on operations codifying safety norms and established best practices, etc.). Of particular importance is a legal recognition of a difference between space objects considered as valuable assets by their owners, and space debris that have no value but may in fact be latent sources of future liability.

- A space traffic management regime has to consider the question of whether harmonizing national space legislation (much of which has yet to be established) and establishing mutually recognized national licensing standards and procedures are essential, since they may provide the building blocks for assuring technical safety. Without minimum, shared regulatory standards future space actors could seek out "flags of convenience" with deleterious effect on sustainability of the domain.

2.5   THE IMPORTANCE OF A SUITABLE LEGAL FRAMEWORK

Effective legal and operational constructs are critically important, especially in missions that require the pooling or sharing of operational data. An effective legal framework is readily understood and enforceable. It incentivises desired behaviours for stakeholders from multiple jurisdictions while discouraging unsuitable behaviours. While liability for space actions – per the Outer Space Treaty – ultimately rests with the launching State(s), many non-State satellite operators take a pragmatic private-law (i.e., contractual agreement) approach. Contractual agreements may be used to define a minimum acceptable level of performance, procedural matters, and critical business and financial terms. This includes pricing; the protection, use, and dissemination of non-public and derived data; dispute resolution mechanisms; data contribution requirements; choice of law and jurisdiction; and intellectual property rights. Finally, contractual approaches can address the complex risk allocation and liability issues to a precision not addressed in contemporary formal treaties, protocols, and conventions.

2.5.1 Attributes of a globally relevant SSA and STM system

A comprehensive SSA and STM system can enhance safe and sustainable conduct of space activities, incorporating international standards, guidelines, multilateral data exchange, registration, notification, and coordination of launch, on-orbit, reentry, safety, and environmental events. To provide decision-quality services, such a system must combine government, satellite operator, and commercial SSA data at the observational level; freely provide a basic level of service to spacecraft operators while not adversely harming established commercial SSA and STM avenues; appropriately protect intellectual property and proprietary data issues associated with international governmental military, civil, and commercial operator space data; apply advanced algorithms and SSA hardware; have high availability; be transparent; and adopt international space data message standards and operational standards to be relevant to the global space market. Ideally, a comprehensive set of evaluation criteria [2] should be adopted to help potential users of such systems be able to discriminate between good and bad SSA and STM systems.

To provide decision-quality services, an STM system must be based upon a robust framework exemplified in Fig. 2.1. An overseeing, decision-making authority manages the entire process, which includes the required quantities and qualities of optical, radar, and passive RF data; incorporation of space operator data; advanced analytics and data fusion; and the most up-to-date space weather history and predictions.

Space operators have a wealth of authoritative information that they may be willing to share with others in the interest of space safety. The upper

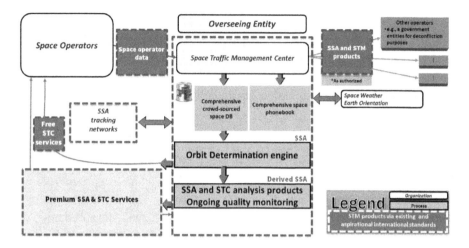

Figure 2.1: Major components of a comprehensive STM system

left box in Fig. 2.1[1] depicts data from contributing space operators, whether they be operating satellites, launch booster and upper-stage vehicles, sub-orbital/exoatmospheric vehicles (e.g. space tourism), high-altitude balloons, or airships. Operator space platforms may host sensors and systems that can provide valuable in-situ measurements of orbital debris, spacecraft charging, and space weather proxies to aid the development and tuning of refined orbital debris models, space weather predictions and models, and dynamically calibrated atmosphere models. In this construct, contributing operators are encouraged to report any satellite and launch vehicle anomalies [12] they experience in the interests of a shared understanding of space risk.

As shown in the upper right-hand corner of Fig. 2.1, this concept departs from other SSA and STM concepts by explicitly recognizing that there will always be nations and/or organizations who want to do their part to minimize space debris but are unwilling or unable to participate directly in a regional or global STM system due to national security or commercial concerns. Such entities are called "Willing Non-Contributor Operators" (WNCOs). When authorized by the original data owners, SSA and STM data can be shared with them according to a fine-grained user access methodology. WNCOs can use this decision-quality information to screen against their non-public (due to intellectual property or classification reasons) space objects to preclude collision or RFI events from occurring.

---

[1] For hardcopy book readers: colour versions of the greyscale figures contained in this chapter can be downloaded from the book product page on the book product website. Please see the preface of this book for more information.

If one or more WNCOs sufficiently trust the STM operator, the STM framework can be optimized further by integrating the public-facing STM and SSA services with the WNCO's. This allows authoritative public data to be merged with WNCO data in a self-consistent manner, greatly reducing transmission bandwidth and latencies and eliminating the need to normalize data between systems.

2.6   HOW ARE CONJUNCTIONS ASSESSED?

Potential collision threats are identified by the SSA system in a process called Conjunction Assessment in the manner shown in Fig. 2.2. The SSA system's aggregate network of sensors tracks all the objects it can. Measurements, or "observations," regarding each space object are sent to observation association (OA) and orbit determination (OD) processing engines. Advanced OD systems can also ingest data about the object's planned manoeuvres when they are provided; if not, the SSA system can still detect, characterize, and account for any manoeuvres that were performed.

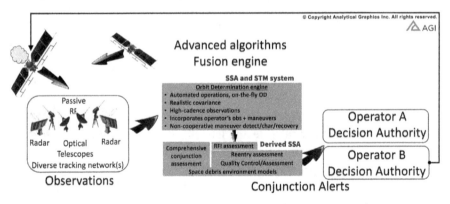

Figure 2.2: The conjunction assessment process begins with SSA sensor observations being collected, from which orbits are determined and potential collision risks are identified. Operators are notified of the collision threat via a Conjunction Data Message. The spacecraft operator can mitigate the impending collision threat

Automated OD algorithms solve for the orbits of all tracked objects and predict their position and position uncertainty into the future. This predicted information is fed into the conjunction assessment process alongside error metrics and space object metadata so that the process can provide an alert if any tracked objects come so close to one another as to exceed an operator's warning threshold.

There are many different types of warning thresholds, ranging from straightforward (predicted miss distance) to somewhat complex collision probability incorporating the uncertainty in the predicted orbits and information about the shape and orientation of the objects involved. An operator's choice of threshold type may be driven by crew resources, available data, and the orbit regime their spacecraft occupies.

In many cases the SSA data required to evaluate such complex metrics is unavailable. Specifically, space object dimensions or overall length, flight attitude rules, and realistic error metrics for supplied SSA positional predictions are often unavailable. Unfortunately, the operators' avoidance manoeuvre go/no-go criteria require these inputs and are typically quite sensitive to any input errors. Many contemporary SSA systems simply assume default values for these parameters without informing the SSA recipient of what those default values are.

When a conjunction is identified, the operator works with the SSA and/or STM service provider to determine if an avoidance manoeuvre needs to be conducted and, if it does, what avoidance strategy to use. The operator uploads the proper commands, the spacecraft manoeuvres, and if all is completed successfully, the two spacecraft pass unhindered.

The situation is further complicated if the second space object is debris, as the U.S. currently does not provide an assessment of object size and positional uncertainty is largely unavailable.

Compounding this complexity, SSA and STM are not one-size-fits-all because the threat profile and the timeliness, completeness, accuracy, and transparency of available SSA data are highly dependent upon the orbit regime. Spacecraft operators in less-dense orbital regimes may have the luxury of being overly careful and manoeuvring whenever another object comes remotely close because they have sufficient fuel margin to ensure safety. Conversely, operators in high-density orbital regimes will not have the luxury to avoid everything that comes remotely close because the millions of potential close approaches would rapidly deplete their staffing resources and fuel budgets.

The safety thresholds that an operator selects and employs tend to be driven by spacecraft cost, mission priority, perceived value to their customer, potential value of derived data, and how long it takes to replace the mission capability by another means. In stark contrast, a spacefaring country (a "State Actor") likely decides to regulate the safety thresholds, algorithms, and metrics employed to be consistent with internationally-adopted treaties, principles, and guidelines designed to ensure the long-term sustainability of the space environment.

Fortunately, spacecraft operators are actively participating in LTS-promoting entities such as the Space Safety Coalition [13], the Space Data Association [14] (SDA, now in its tenth year of commercial space operator self-funded SSA and Space Traffic Coordination services), the World Economic Forum's Space Sustainability Rating (SSR), and a number of internal space safety initiatives.

2.7 WHAT MAKES SSA AND STM SO CHALLENGING?

Accurate, comprehensive, and timely space situational awareness is foundational to space traffic management, yet few appreciate the many moving parts required to obtain such SSA. As shown in Fig. 2.3, major sections of the SSA chain include at least seven components: the SSA system itself, the sensors that observe the space situation, the data-pooling and fusion engine, SSA analytical and algorithmic foundation, all of the data associated with space objects, the orbit determination and prediction tools, and RFI analysis tools. These major components underpin all SSA, STM, and regulatory approaches.

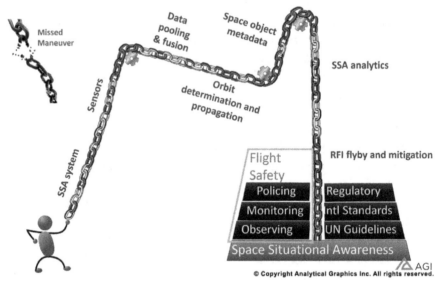

Figure 2.3: All the components of a comprehensive SSA and STM system

Disconcertingly, a failure in any of the many links in this chain can invalidate SSA. The collective performance of this entire functional chain determines whether SSA products are useful; in space safety and sustainability, there is no such thing as partial credit for getting it "mostly correct". This was the case when the primary SSA systems failed to identify the 2009 Iridium/COSMOS collision because a single Iridium stationkeeping manoeuvre was omitted from the SSA threat-monitoring processes. This omission triggered an event whose pre-manoeuvre probability of occurrence was estimated as one in one trillion-trillion-trillion. If that manoeuvre had been incorporated, the estimated probability would have jumped to one in one thousand (and subsequently skyrocketed all the way to a value of one once the collision occurred).

Having led the development of the first U.S. probability-based Launch Collision Avoidance (LCOLA) system in 1996, we understand how difficult it is to assemble and validate all of the requisite links of this SSA chain. It can be too easy to celebrate the achievement of building this end-to-end process, at the expense of how much further one must go to ensure that inputs, algorithms, and data products are sufficiently complete and accurate to support decision-makers and ensure flight safety. It is unsettling that many of our current SSA processes do not have effective quality control mechanisms, and it is too tempting to simply assume that an end-to-end process works without actively monitoring its accuracy and usability.

Some of the barriers to achieving this end-to-end SSA process are listed in Fig. 2.4, wherein every listed aspect has the potential to render an SSA-based prediction invalid.

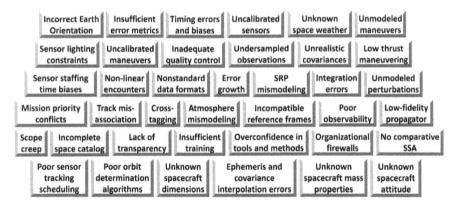

Figure 2.4: Modelling challenges that can trip up SSA systems

Spacecraft operators and countries have limited resources to develop, operate, avoid collisions, and communicate the importance of flight safety to management. New Space entrants often lack access to SSA and STM subject matter and expertise, making it especially difficult to understand the nuances and ramifications of spaceflight safety. Without that expertise it is difficult to convey two crucial points to high-level decision-makers and funding authorities: the criticality and cost benefits of paying for actionable SSA products to help ensure flight safety, and the adverse impacts to customer base and brand recognition if space assets are not adequately protected.

2.8 WHO PROVIDES SSA AND STM SERVICES?

In truth, it is a stretch to assert that any one political entity provides comprehensive STM services today, as no coordinated body manages, controls, or directs operators' spacecraft. But SSA and STM services are already available today through U.S. domestic and foreign entities and in the global commercial marketplace. These services could readily support the provision of STM services needed for space sustainability.

Although some advocate for a single entity to provide global SSA and STM services [15] (typically based upon the International Civil Aviation Organization or ICAO model for air traffic control), there have historically been only a handful of nation-states and commercial entities possessing the resources, technical means, and global reach to effectively maintain SSA. Legacy provision of SSA and STM services has typically been provided by the United States government. However, increasingly, foreign governments and commercial providers are stepping up to furnish enhanced SSA and STM services.

2.8.1 Legacy SSA and STM services

SSA and STM services have long been provided free of charge to the spacecraft operator community through the U.S. Combined Space Operations Center (CSpOC) by the 18th Space Control Squadron (18SpCS). Note that the Department of Defense (DoD) is currently reorganizing around space as an Area of Responsibility (AoR), resulting in the changing/updating of unit and asset nomenclature. Using data from the Space Surveillance Network (SSN) – a global collection of ground-based radar and optical sensors and space-based optical sensors (Fig. 2.5 [16] and Fig. 2.6 [17]) – the DoD does a laudable job of providing global SSA services (a complex process that requires Congressional authority), the institution of requisite operational procedures, and the construction and maintenance of partnerships with various foreign government and commercial entities. The DoD should be commended for its foresight in supporting sustainable spaceflight and its diligence in establishing an SSA-sharing paradigm.

However, as acknowledged by the CSpOC, these assessments are purely intended as a "heads-up" of potential collision risks rather than a full-scale

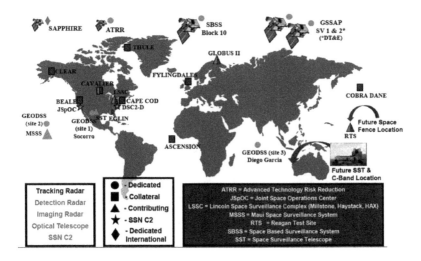

Figure 2.5: Space Surveillance Network configuration

conjunction characterization suitable for collision avoidance decisions. Today's U.S.-provided legacy capabilities, as realized in the SSA Data Sharing agreements and provided on space-track.org, are not always adequate to generate operationally-relevant, decision-quality SSA information with sufficient accuracy, timeliness/responsiveness, capacity, and unambiguity that space operators require.

Fault does not lie with the men and women in uniform performing this duty. Rather, the problem lies with the tools with which they have been provided, which were designed for the vastly different space operational environment of the 1970s and 1980s. Forty to fifty years ago, space was not considered a contested domain. Operators had a "space is big" mentality (i.e., with relatively few objects in orbit, the collision risk was acceptably small). These tools supported the goals of this early stage of the space age, which was simply to be able to "maintain custody" of (i.e., reacquire and monitor) space objects. But in the dramatically evolved space operational environment of 2020 – a substantially increasing amount of debris, new technology capable of maintaining this debris for collision avoidance warning, and the advent of large constellations – these legacy tools are unable to achieve the SSA and STM performance levels required to ensure spaceflight safety. Unfortunately, the mere existence of, and familiarization with, a process based solely on this imperfect database has lulled many operators into a false sense of preparation and security – some see this system as "good enough," despite its obvious shortcomings.

Figure 2.6: U.S. radar in Thule, Greenland

Space was not previously considered a warfighting domain. It is now. The U.S. has gone to great lengths – establishing a new branch of the armed services and a new unified command, U.S. Space Force and U.S. Space Command, respectively – to manage space as a warfighting domain. U.S. Space Command (formerly U.S. Strategic Command) openly advocates that the responsibilities of SSA sharing and the provision of spaceflight safety services to civil and commercial entities should fall outside the DoD to allow warfighters to focus on national security issues.

While the DoD has provided a commendable public service by establishing and operating their free collision warning service, growing national security space concerns and the increasingly complex space operational environment have challenged the status quo. These services are not sufficiently accurate or robust for generating decision-quality information; it cannot respond to rapidly evolving situations, process all the necessary data, or provide the transparency and availability necessary for widespread international adoption. These shortcomings result in an unacceptably high false positive alarm rate, which unfortunately has pre-conditioned some operators to ignore such collision risk warnings. As such, this "free" service does, in fact, come at a cost – namely, excessive (and unknown) risk acceptance.

Unfortunately, the burden of significant legacy infrastructure and acquisition culture has made it difficult for the U.S. DoD to modernize its SSA capabilities. The U.S. Air Force has spent more than $3B over the last 30 years in failed attempts to modernize its space command and control infrastructure (including SSA), yet it is still reliant upon decades-old technology.

Additionally, although total collision risk across the entire space population is significant and about to substantially increase with the introduction of large Low Earth Orbit (LEO) constellations, collision risks may seem negligible to the individual satellite operator. Satellite operators may underestimate these risks and exaggerate the measures they've taken to address them when unduly influenced by diverse financial, regulatory, competitive, and cultural pressures. Similar to other tragedy-of-the-commons situations, it is understandable (but still unacceptable) that many operators' business models fail to incorporate environmental worries. Indeed, the decreasing satellite manufacturing costs from mass production and miniaturization may already preclude "natural" market forces from motivating operators to protect the shared satellite operations environment. From a purely financial perspective an individual operator may be willing to risk losing a satellite to a collision, especially for small/non-economic (e.g., academic) operators or large constellations with multiple redundancies and quick re-launch/refurbish capabilities.

These considerations, coupled with a false sense that our operational space is an infinite expanse, leads some satellite operators to unilaterally accept their collision risk on behalf of the entire space community. They may rely on the inadequate, free legacy services to justify collisions – after all, how can a United States Government-provided service be insufficient? Yet collisions, once they occur, are irreversible. They can potentially degrade the global space economy's operational environment, causing long-term costly effects for the rest of the space operator community.

An old adage asserts that one should not look a gift horse in the mouth. However, doing so with "free-to-use" STM systems is an entirely different matter. Today's premier SSA system operated by the U.S. Space Command does not charge non-governmental users for services rendered. Interestingly, this leads to conditions where their service and data are offered on an as-is basis to all. There are multiple benefits to leveraging the significant public investments made in establishing and operating the SSN and using its defence-oriented capabilities to provide ancillary services to civilian and commercial space operators. These services are not, however, those between a customer and a service provider, but rather between a State and its supplicant users. Users may request changes but lack effective means to cause changes; hold the United States Government accountable for non-timely, missing or under-performing issues; or gain realistic insight into the nature and quality of the underlying dual-use systems supporting activities (e.g., missile warning sensor performance).

For critics of the current system, the real issues come after possible colli-sions are identified. Once a warning message is sent the military has no further role in what happens next – to ensure that people are making decisions that benefit the entire space community.

2.8.2 Other global SSA and STC providers

Other countries operate SSA systems but their products are not currently as widely accessible. However, several counties have recently advanced con-certed efforts to build and assemble stand-alone SSA systems, in part to pro-vide SSA and STM information and services in support of spaceflight safety and space sustainability. Most notable to date is the European Union's Space Surveillance and Tracking System (EUSST).

In 2014, the EU decided to set up the EUSST via a consortium of EU member states, networking their national SSA operations centres and sen-sor (ground-based radar, telescope, and laser station) capabilities together. The consortium initially consisted of France, Germany, Italy, Spain, and the United Kingdom and has since expanded to include Poland, Romania, and Portugal (it remains unknown how Brexit will affect United Kingdom partic-ipation). The consortium works in cooperation with the EU Satellite Center (EU SatCen) to provide conjunction analysis, reentry analysis, and in-orbit fragmentation information and services to the EU, member states, and other registered users.

Figure 2.7: GRAVES transmitter (left) and receiver (right)

Sensor elements of the EUSST notionally include France's Grand Réseau Adapté à la Veille Spatiale (GRAVES) radar [18] (Fig. 2.7) for LEO tracking and the Télescopes à Action Rapide pour les Objets Transitoires (TAROT) telescopes [19] (Fig. 2.8) for GEO tracking, as well as other devoted or col-lateral tracking radars (e.g., SATAM, ARMOR 1&2, and NORMANDIE). Germany employs the Fraunhofer Institute for High Frequency Physics and Radar Techniques (FHR) Tracking and Imaging Radar (TIRA) [20] (Fig. 2.9a) for the observation of space objects as well as for the characterization of the small particle debris environment in low Earth orbit. The United Kingdom operates the Chilbolton radar [21] (Fig. 2.9b) with LEO tracking capability and the Starbrook optical sensor for GEO tracking. This is not an all-inclusive description of the full sensor capabilities of the EUSST, and the newer members are sure to add additional capabilities.

Figure 2.8: TAROT telescopes

The various consortium members' space operations centres share in the necessary data ingestion and processing to generate the information that is then made available through the EU SatCen. And each of the consortium members continue to individually advance/upgrade their national capabilities as well as their space policies with respect to SSA and STM.

Russia has a system that is similar to the SSN but covers different regions of space. The International Scientific Observation Network (ISON) of telescopes provides a detailed catalogue of objects in geostationary orbit. China also employs SSA capabilities, but there is little transparency into the scope of those capabilities or their applications outside of national purposes.

In addition, numerous other countries are beginning to implement rudimentary SSA capabilities with an eye on STM. For instance, Japan is in the processing of developing an SSA system, in part to support STM, and exploring the possibility of augmenting their national capability with commercial capabilities. India is working to establish SSA capabilities to support their upcoming human spaceflight launch and operations, scheduled to begin in 2022. New Zealand recently established a demonstration capability with a commercial radar data provider to explore SSA and STM concepts.

Figure 2.9: (a) FHR TIRA radar. (b) Chilbolton radar

As these nations, and others, continue to move forward, they also work to update their SSA and STM policies and strategies and work within the various international fora, as well as through bi-lateral and multi-lateral arrangements, to pursue cooperative efforts with regard to SSA and STM.

2.8.3   Defining the commercial SSA and STM option

Investment in SSA and STM no longer has a direct relationship with SSA and STM capability. Just as third-world countries have installed advanced cell-based telephony systems at a fraction of the cost of using copper-based technologies, a favourable combination of increased capacities, capabilities, and performance at lower cost has enabled a number of competing commercial SSA-system alternatives to emerge. Already, several SSA entities are fully operational (Technology Readiness Level 9), offering comprehensive SSA data and services to the space operator community. It can be difficult on the surface to determine which of these entities are capable of meeting a space operator's stringent needs. Operators are typically looking for a well-vetted, transparent, fully operational SSA system with high availability, advanced algorithms, automated processing, assured availability, and a secure and trusted computational framework.

Similar to trends in reusable launches, active debris removal, remote sensing, and communications, commercial ventures anticipate SSA needs and accept development risks up front. Commercial ventures leverage modern computing techniques, algorithms, and technology to deliver and operate innovative SSA capabilities meeting today's space operational environment challenges. For instance, commercial enterprises, leveraging affordable but more advanced, technology for ground-based sensors, have installed several hundred sensors globally – far exceeding the numbers of sensors maintained by national governments. In contrast, there are fewer than twenty ground-based sensor sites in the U.S. Space Force Space Surveillance Network.

Commercial companies establish a cycle of innovation to promote and support continual improvements, thus motivating the commercial marketplace to seek their services. Leveraging cost effectiveness through commercial approaches makes for affordable investment in efforts to modernize SSA capabilities. It is precisely this innovation and cost effectiveness that allows countries who are new entrants to SSA to transition to providing a capable SSA service quickly.

To compete with government-provided or commercial competition alternatives, privately-funded commercial STM solutions must focus on efficient means to maintain a predefined level of service quality for their customers or risk losing them. Service quality, evolution, and accountable behaviours are hallmarks of these commercial enterprises.

2.8.4   History of SSA and STM commercial service options

The commercial community's involvement in SSA began in 1985, when Dr. T.S. Kelso created the first public space data portal, CelesTrak. The Satellite Orbital Conjunction Reports Assessing Threatening Encounters in Space (SOCRATES) online conjunction assessment tool was added to CelesTrak in 2004. Originally based on the USAF's lower-precision orbit theory (SGP) for all space objects, SOCRATES was upgraded in 2008 to directly ingest highly accurate, operator-predicted spacecraft positional information incorporating their planned spacecraft manoeuvres.

SOCRATES then led to the space-operator-formed Space Data Association (SDA) in 2009, which for over a decade has provided safety-of-flight services to the global space operator community. The SDA was formed to unite civil, commercial, and military operators in supporting the controlled, reliable, efficient, and secure exchange of data needed for flight safety and satellite operation integrity. The SDA itself operates on a nonprofit basis, has a legal structure, and is incorporated in the Isle of Man. Its legal structure allows the protection and enforcement mechanisms to ensure that all data provided to the SDA are protected as needed.

The SDA has operated its cloud-hosted Space Data Center (SDC) for over a decade now, pioneering many of the traits now widely accepted as baseline requirements for a capable STM system. Among those baseline requirements are a robust legal framework, military-grade computational security, geographically diverse processes, very high availability, ongoing forensics, comparative SSA analyses, and format-agnostic ingestion of operator data with machine-to-machine interfaces and verified data normalization converters. The SDC has evolved to be one of the largest clearinghouses for spacecraft operator data in the world. Since its inception, it has provided an operator phonebook that is sufficiently granular by area of responsibility, location, and management level to allow efficient operator communication. The association uses comparative SSA for ongoing quality control and to identify discrepancies, allowing SDC analysts to alert spacecraft operators, governmental SSA centres, and commercial SSA of any observed discrepancies and/or high-risk collision threats.

The 30 global space operators currently participating in the SDC operate spacecraft spanning all orbital regimes, form factors, and mission types. Safety-of-flight analyses are currently performed for 786 spacecraft (513 spacecraft in LEO and MEO and 273 spacecraft in GEO). The SDC's innate ability to crowd-source space data from spacecraft operators and merge that with accurate space debris catalogues from the U.S. Air Force's space catalogue in both SGP and Special Perturbations (SP) formats has allowed the SDC to generate decision-quality STM analytics. The SDC also serves as a distribution hub for space data, a focal point for comparative SSA and quality control, and a high-availability provider of SSA and STM services. This decade of SDC operation has confirmed that rather than continuing to ask our community's long-standing question, "Is my SSA data better than yours?", the SDC system embodies the notion that decision-quality SSA and STM are only achievable by the use of advanced algorithms, assured processing, and the aggregation and fusion of all-source data.

More recently, as many as 14 commercial SSA service providers have formed, half of which are U.S. companies.

2.8.5 Characterizing SSA system performance

Ideally, one should try to characterize absolute positional accuracy (a primary SSA metric) as a function of time. Unfortunately, there are few publicly available, positionally well-known reference or "truth" objects in space so it is difficult to draw statistically relevant conclusions on SSA system performance from that small set of objects. Since accuracy is a combination of system biases and the inherent repeatability (or precision) of an SSA system's predictions, system accuracy can instead be bounded by estimating that system's precision over a large data set. Any observed imprecisions are typically caused by insufficient SSA force models, unknown or modelled events (e.g., unknown space weather or unknown manoeuvres), undersampled observations, and/or algorithmic or process-based SSA deficiencies.

One can statistically characterize the repeatability of predicted positions over a large set of objects and sufficiently long time span. As an example, consider the characterization [22] of the legacy SSA system operated by the 18th Space Control Squadron. One can statistically assess the precision of the entire set of shared Special Perturbations (SP, 18SPCS high accuracy) catalogue of 17,958 Resident Space Objects (RSOs) through recurrent positional differences of each RSO's sequential ephemeris. For collision avoidance, such precision statistics associated with orbit prediction time spans of between one and two days were of greatest interest because that prediction time is most relevant to an operator's typical Observe/Orient/Decide/Act (OODA) loop for conducting collision avoidance manoeuvres.

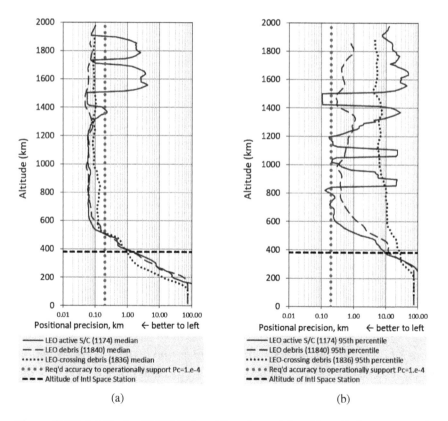

(a)   (b)

Figure 2.10: (a) Typical LEO SP positional precision (1- to 2-day propagation, 50th percentile) (b) 95th percentile LEO SP positional precision (1- to 2-day propagation)

The median and 95th percentile statistical discrepancies in the precision (repeatability) of one- to two-day positional predictions spanning the entire range of true anomaly (0° – 360°) are characterized for LEO (0 – 2,000 km altitude) in Figs. 2.10a and 2.10b, and for GEO in Figs. 2.11a and 2.11b. Release of these figures has been authorized by United States STRATCOM and the 18th Space Control Squadron. The straight vertical dotted line denotes the accuracy required to operationally support the collision probability threshold of 0.0001 commonly used by spacecraft operators as a collision avoidance manoeuvre Go/No-Go criterion.

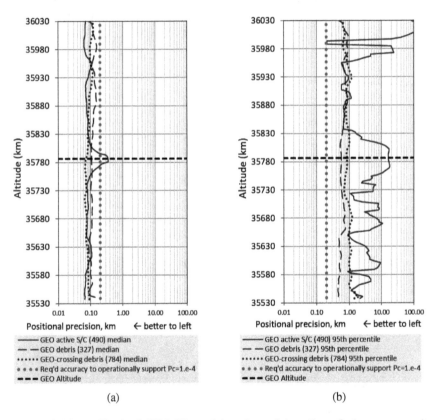

(a)   (b)

Figure 2.11: (a) Typical GEO SP positional precision (1- to 2-day propagation, 50th percentile) (b) 95th percentile GEO SP positional precision (1- to 2-day propagation)

Note that while typical, or 50th percentile, SP ephemeris precision often meets (i.e., is on the left-hand side of) this limiting accuracy threshold, there are altitude ranges, orbit types, and manoeuvrability categories for which SP performance fails to meet the threshold. When one further considers higher levels of occurrence, or 95th percentile, this limiting accuracy threshold may often be exceeded for certain orbit regimes (e.g., space weather below 700 km and high-eccentricity orbits) and object types (e.g., active, manoeuvring satellites).

2.9 SSA AND STM AS THE FOUNDATION OF LONG-TERM SUS-TAINABILITY OF THE SPACE ENVIRONMENT

The basic constituents of space sustainability are clear. We must prevent predictable collisions (prevention), minimize the creation of new debris (mitigation), and remove massive derelict LEO objects (remediation), as shown in Fig. 2.12a. If space were a school playground, the rules would be "don't hit each other, play nicely, don't litter, and put your toys away." The relative size of each constituent piece is meant to notionally convey the authors' view of its relative importance in the overall scheme of ensuring long-term sustainability. What has become increasingly evident based upon space population long-term evolution studies is that successfully disposing of spacecraft and mission-released debris once their mission is completed is the most important step we can take.

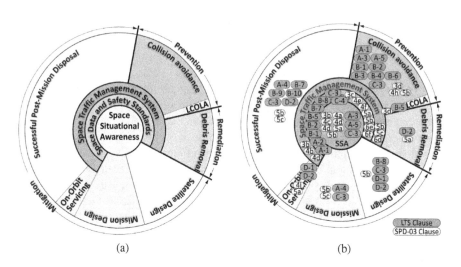

Figure 2.12: (a) SSA and STM are foundational to all long-term sustainability of space activities (b) Mapping of the 21 LTS Guidelines and U.S. SPD-3 clauses to the LTS block diagram

Each of these constituent pieces comprises one or more mission life cycle aspects. All of them ultimately depend upon SSA and STM.

2.10 THE IMPORTANCE OF DECISION-QUALITY SSA AND STM SERVICES

As stated previously, it is already challenging to develop and integrate all of the requisite tools and capabilities required to provide SSA and STM services. It is all the more difficult to ensure that those services yield accurate, timely, comprehensive, and transparent metrics that spacecraft operators can depend on when deciding whether to avoid a collision threat. Where collision avoidance is concerned, launch and spacecraft operators can take one of four paths:

(1) Don't believe that collisions are a credible threat, so no SSA monitoring or collision avoidance action is taken;

(2) Agree that a credible threat exists but don't believe that the SSA and STM results they are provided are credible, so no avoidance action is taken;

(3) Agree that a credible threat exists and that SSA and STM results have known deficiencies, so they apply large margins of safety to SSA and STM results to compensate for those deficiencies; and

(4) Agree that a credible threat exists and tirelessly seek the most accurate, timely, comprehensive, and transparent SSA and STM services to support safety decision-making processes.

It is critical to realize that where human spaceflight is concerned, Categories (1) and (2) are entirely unacceptable: one must always ensure the safety of human-carrying and human-habitable space objects. The risk of colliding with them must never be dismissed whenever overlapping orbit altitudes could even remotely place humans at risk. Conversely, Categories (3) and (4) may be acceptable so long as human safety is assured through human safety collision avoidance. Once human safety is assured, it's worth examining "mission assurance" collision avoidance separately for both the launch and on-orbit context for each of these four Categories.

For launch mission assurance collision avoidance, the typically short duration of flight means that collision with the on-orbit population is not likely. Parametric examination of collision risk across all launch times for the ascent and early orbit trajectories being flown may tempt one to conclude that collisions will not occur during the launch phase and that Category (1) is justifiable. But do not be misled by statistics; asserting that the median (50th percentile) collision probability for launch justifies Category (1) is akin to concluding that because the median distance between your car and oncoming traffic is two meters, you needn't worry about oncoming traffic coming into your lane.

For on-orbit mission assurance collision avoidance, research has identified a substantive level of collision risk in both LEO and GEO regimes. Once again, Category (1) and Category (2) are unacceptable; the space community must work tirelessly in a comprehensive, leave-no-stone-unturned effort to improve SSA and STM to provide credible SSA and STM services.

For on-orbit mission assurance collision avoidance, Category (3) is a path that may be justified if the following conditions are met:

- the spacecraft has ample manoeuvring fuel;

- the operator has ample flight dynamics staff resources; and

- conjunctions are rare (e.g., the spacecraft is placed in an orbital regime in which conjunctions are infrequent).

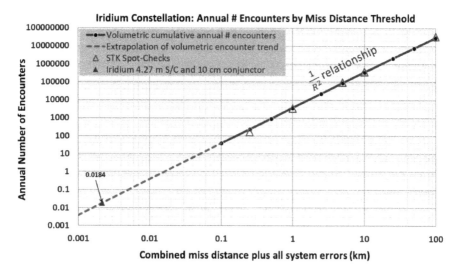

Figure 2.13: The number of potential threats operators must process depends exclusively upon how accurate SSA predictions are

But as the number of active spacecraft and debris fragments increases and our tracking of space objects improves, the number of potential collision threats is growing to the point where operators no longer have the sufficient fuel, staffing, or low collision risk required for Category (3) to be feasible. This is graphically depicted in Fig. 2.13 which illustrates how the number of potential threats that operators must process depends almost exclusively upon how accurate the SSA data is. Increased accuracy obtainable from advanced SSA algorithms can substantially reduce spacecraft operator workload by eliminating numerous false alarms [23]. In a roughly homogenous space environment (as exists in the higher density mid-LEO region between approximately 700 – 900 km altitude), encounter rates adhere quite well to the inverse radius-squared relationship codified in the kinetic gas theory. This is important because it is the key to how the global space community can address safety-of-flight in the New Space era. To use a William Tell analogy, how close one is willing to come to a target (or, in this case, conjunct debris fragment) depends upon the accuracy of the marksman. If the underlying SSA system can be trusted to provide warning of conjunctions accurately with depictions of uncertainty that are realistically small, it's more justifiable to stand one's ground.

2.11 THE IMPORTANCE OF STANDARDS IN SSA AND STM

As identified in both the UN COPUOS LTS guidelines and U.S. Space Policy Directive 3, international standards should underpin the data pooling and/or exchange with any SSA and STM framework. It is vital that our space community have a shared understanding of the role and importance of international standards.

International standardization helps prevent technical nationalism and provides a common reference framework in a commonly accessible language to facilitate trade and technology transfer between global space actors. Standards describe performance requirements and interfaces in verifiable, achievable ways that are suitable for incorporation into contractual mechanisms. Standards also promote capacity building and the sharing of technical knowledge.

Some mistakenly believe that standards stifle innovation and potential; the reality is precisely the opposite. Standards are living documents that are periodically reviewed to ensure marketplace relevance, technical currency, and completeness. ISO standards typically require two to three years to develop. Once published, each standard is reviewed every five years to determine whether it should be renewed, revised, or retired. Standardizing recurring activities can be tremendous market and innovation enablers.

Within the ISO governance structure, the technical committees and their subcommittees and working groups develop and maintain the standards. ISO comprises 245 such technical committees containing over 100,000 global subject matter experts. ISO currently has 22,000 active international standards published in English, French, and Russian. Our focus will be on those intended to develop standards for space activities.

Also established in 1947, ISO Technical Committee 20 (ISO/TC20) is one of the most prolific ISO technical committees in international standardization. With over 600 published standards developed under the broad umbrella of the committee and its subcommittees, ISO/TC 20 maintains a significant, relevant presence in the aerospace industry. Within TC20, two subcommittees are developing space standards – Sub-Committees 13 (SC13) and 14 (SC14).

ISO TC20/SC13 develops international space data message standards. SC13, functionally equivalent to CCSDS, comprises 11 space agencies globally. Space data message standards assembled by its Navigation Working Group are particularly relevant to the long-term sustainability of space. These standards concern ways to exchange space data, including such data types as *attitude, conjunction, event, orbit, pointing, reentry*, and *tracking* data. The Orbit Data Message (ODM) is the most widely downloaded Navigation Working Group standard today. Additional standards are needed to address *anomaly, fragmentation, geolocation, launch, RFI, RF characteristics*, and *Rendezvous Proximity Operations/Satellite Servicing Operations (RPO/SSO)* events.

ISO TC20/SC14 develops best practice standards for space systems and operations. All disciplines of SC14's seven working groups are relevant to the long-term sustainability of space activities.

2.12 WHERE STANDARDS FIT WITHIN THE SSA AND STM LAND-SCAPE

International standards are essential enablers in the context of the SSA and STM landscape. Returning to the overall STM framework depicted in Fig. 2.1, one can insert space data exchange standards into this framework as shown in Fig. 2.14. In this figure, the boxes with curved corners represent organizations, boxes that have dark grey dashed lines represent internationally-standardized space data messages, and other boxes represent processes and analyses. As in Fig. 2.1, the rectangles on the upper right represent external "Willing Non-Contributor" operators.

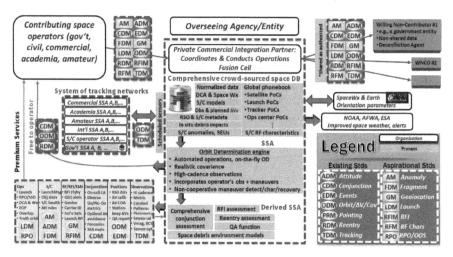

Figure 2.14: The important role of space standards in a comprehensive STM system

The box in the upper middle of Fig. 2.14 depicts space data aggregated from contributing operators of spacecraft, launch booster and upper stage vehicles, sub-orbital/exoatmospheric vehicles (e.g. space tourism), high-altitude balloons, or air-ships. Such organizations have a wealth of authoritative information that they may be willing to share with others in the interest of space safety.

SSA tracking data, refined space weather predictions, and debris campaigns can be shared via internationally-standardized navigation messages that are fed into the STM system. By ingesting space catalogue observations and ephemerides and com-bining that data with actionable operator-provided data, comprehensive, actionable, and timely SSA and STM assessments can be made. Comprehensive SSA and STM results can then be shared back to the space operators, again relying heavily on the current and future internationally-standardized messages, both Conjunction, Orbit,

Fragmentation, and Reentry Data Messages (CDMs, ODMs, FDMs, and RDMs) at a basic level of service, and then as advanced services shown in the box at lower left. Willing Non-Contributor Operators (WNCOs) can also benefit tremendously from the use of internationally-standardized data messages.

2.13   THE EXTREME IMPORTANCE OF SPACE DATA EXCHANGE

For SSA and STM to be as accurate as possible, it is critical that all space actors, space object tracking entities, space weather observers and modellers, and SSA software and algorithm developers pool their resources and expertise. The exchange, aggregation, and fusion of all available data using advanced algorithms and analytics provides the best possible understanding of current and future threats to long-term sustainability. Space data exchange has been perhaps best accomplished using a space data lake model as implemented by the Space Data Association, which allows spacecraft operators and SSA service providers to post their raw data to a computationally and legally-secure framework which then normalizes that data and conducts flight safety assessments without disseminating proprietary or intellectual property data unless explicitly approved by the data owner.

To understand why this is so, consider the primary processes underlying SSA for every space object: orbit determination and orbit prediction. OD is the discipline of estimating an object's orbit based upon past measurements. Once the orbit has been estimated, astrodynamicists employ the mathematical concept of numerical integration to predict future space object positions required for SSA and STM. Numerical integration of a space object's position must incorporate our knowledge of both the object's estimated orbital position at a time t_0, as well as our estimation of accelerations (as a function of estimated position and velocity) at prescribed future times t_1, $t_2, t_3, \ldots t_n$, to obtain the estimated position of the space object at time t_n.

Put into layman's terms, OD is the process of fitting a presumed force model to historical observations, and orbit prediction is the extrapolation of that model forward in time. But as all engineers know, extrapolation can be dangerous.

To understand why extrapolation is dangerous, imagine that you are provided with accurate Global Navigation Satellite System (GNSS) positions of an aircraft every second over a one-hour period. You are asked to (i) determine where that aircraft is at the end of that hour and (ii) predict where that aircraft will be located after another hour has elapsed. Unless you had access to the pilot's flight plan, these are impossible tasks. You have no insight into the underlying air traffic flight rules, the pilot's goals and ultimate destination, steering control yoke inputs and corresponding time-dynamic accelerations, the aircraft's response (causality) to those controls, or many other critical details required to make such a prediction.

Yet somehow, we think that orbit prediction is achievable for space objects. While decades of research and progress have been made in developing gravitational models, reference frame definitions, and space weather modelling, the assessment of accelerations at prescribed future times $t_1, t_2, t_3, \ldots t_n$ can be difficult due to an SSA service provider's inability to obtain, know, or predict such information. Conversely, almost all of the following spacecraft parameters and metadata are known to an exquisite detail by the spacecraft operator:

- Future space weather conditions (solar activity and geomagnetic field) that influence drag forces;
- Spacecraft mass;
- Spacecraft sizes, dimensions, and orientations;
- Spacecraft three-dimensional models;
- Spacecraft materials and reflectivity to support attitude inversion and patterns of life evaluations;
- Spacecraft attitude as a function of time, and its associated influence on the drag forces encountered;
- Spacecraft communications frequencies and Radio Frequency directionality to allow passive RF techniques to augment radar and optical observations;
- Spacecraft activity/status (active, operational, dead, sleeping, partially disabled);
- Spacecraft control laws and level of autonomy (does it determine when to move, and to what degree flight dynamics engineers are able to predict what actions the spacecraft will do in the future (or did in the past);
- Spacecraft operator future manoeuvre plans, especially for low-thrust manoeuvres with thrust vector steering control laws or manoeuvres with duty cycles or rapid sequences;
- Unknown momentum dump control sequences and orbit-affecting attitude manoeuvres;
- Contact info for spacecraft operator management, flight dynamics, and RFI personnel.

These aspects directly influence our ability to estimate the aggregate perturbing force accelerations as a function of time, without which it is impossible to propagate orbits and to assess collision risk and its potential impact on the space environment. Some of these aspects, or at least combinations of some of these aspects such as the ballistic coefficient, can be estimated or inferred by the orbit determination process or from Radar Cross Section (RCS) or Visual magnitude (Vmag) measurements. But there can be a huge cost in doing so: such estimates can have very large uncertainties, leading to collision threat identifications and risk characterizations that are erroneous by orders of magnitude.

With the exception of space weather, this information could be directly and authoritatively provided by the spacecraft operators because they are the experts for their spacecraft. By implementing legal and computational constructs that respect and protect proprietary and intellectual property, SSA and STM systems can generate products that are substantially more accurate, thereby reducing both Type I (false positive) and Type II (false negative) safety-of-flight errors.

2.14 ALGORITHMS DO MATTER: WHY THIS TRULY IS ROCKET SCIENCE

When developing new SSA and STM systems, it is easy to mistakenly prioritize hardware (facilities, sensors, computing and network infrastructure, power backup systems, geographic diversity) above algorithms and software. While hardware is indeed important, SSA and STM systems often fail because of a lack of prioritization on algorithms, advanced analytics, diverse metrics and risk portrayals, and quality control and comparative SSA processes.

In the New Space era, we need to prioritize the incorporation of advanced techniques and the development of brand-new algorithms that substantially improve orbit determination accuracy and prediction (e.g., reducing two-sigma errors to less than 40-200m for GEO, 25-75m for LEO). These are the kinds of advanced technologies that need to be incorporated:

- Input data provenance and queuing of sensors and SSA information;

- Insight into how information was generated and the accompanying metadata;

- Automated all-source sensor metric observation fusion (all formats/standards/phenomenologies);

- Advanced sequential filters that incorporate process noise terms to accurately estimate orbits and their accompanying realistic error depictions;

- OD methods to greatly improve responsiveness to detect, process, and characterize manoeuvres in Near Real Time (NRT) – essentially allowing orbit solutions to be updated as soon as post-manoeuvre data is received;

- Trend analysis in order to establish patterns of behaviour/patterns of life;

- Tools to assess adversarial intent;

- Tools identify anomalous situations that can impact operations and conjunction assessment ;

- New comparative SSA and quality control mechanisms to determine and de-weight suspect sensor data;

- Risk analysis and risk acceptance calculations;

- Collision avoidance characterization and decision aid tools; and

- Techniques to model observed collisions and rapidly notify spacecraft operators placed at greatest risk from the resulting debris fields.

Table 2.3: General performance comparison of sensors that can track both space debris and spacecraft

Sensor type	GEO coverage	LEO coverage	Not lighting-dependent	All weather	Range	Range rate	Angles
Monostatic Radar	◒	●	●	●	●	◒	◒
Bistatic Radar	◒	●	●	●	●†	◒	◒
Optical Telescopes	●	◒	○	○	○	○	●
Passive RF (TDOA/FDOA)	●	●	●	●	●†	●	○
LIDAR	●	●	◒	○	●	○	◒

† Derived quantity   ● full   ◒ partial   ○ little or no capability

2.15 ALGORITHMS AND INPUTS: THE CASE FOR COMPREHENSIVE, ALL-SOURCE DATA FUSION

Spacecraft operators are increasingly burdened with collision threats; degrading safety; and draining analyst, management, and manoeuvre fuel resources. This burden can be effectively minimized by ensuring that the predicted positions of space objects are as accurate as possible. Accuracy requires two things: (i) exquisite historical orbit solutions based on based upon past sensor observations and environmental conditions; and (ii) incorporation of best estimates of all relevant future perturbing forces (drag, solar radiation pressure, spacecraft manoeuvres, momentum dumps, etc.) when predicting the evolution of that orbit solution.

Orbit prediction, then, is based on the conglomeration of sensor observations, environmental and space weather predictions, and perturbing forces (including manoeuvres). Sensor observations can be thought of as being *cooperative* (e.g., provided by the spacecraft operator) or *non-cooperative* (independently observed, such as is required to track orbital debris). The general strengths and weaknesses of sensor types capable of tracking space debris and/or non-transmitting spacecraft are depicted in Tab. 2.3. Strengths and weaknesses of sensors requiring spacecraft that transmit radio frequency (RF) energy are shown in Tab. 2.4.

A frequently unexploited way to dramatically improve positional accuracy is by applying advanced orbit determination algorithms that incorporate as wide a mix of sensors, sensor types, and viewing geometries as possible. The benefits of this are shown in Fig. 2.15. The figure compares the two-sigma error *bubble* or ellipsoid around the tracked object's estimated nominal position obtained using only *radar or passive RF* observations, only *optical* telescope observations, or the *combined* set of observations. This illustrates the substantially reduced error ellipsoid that can be obtained when both radar and optical observations are combined or *fused*.

Table 2.4: General performance comparison of sensors that can only track actively-transmitting spacecraft

Sensor type	Doesn't require operator cooperation	GEO coverage	LEO coverage	Not lighting-dependent	All weather	Range	Range rate	Angles
Spacecraft transponder ranging and range rate	○	●	○	●	●	●	●	◐
1-way Doppler	●	●	●	●	●	○	●	○
Radio Telescopes	●	●	●	●	●	○	○	●
Passive RF (TDOA/FDOA)	●	●	●	●	●	●†	●	○
Onboard GNSS	○	●	●	●	●	●†	●†	●†

† derived quantity   ● full   ◐ partial   ○ little or no capability

Similarly, Fig. 2.16 shows how errors can be dramatically reduced by fusing operator ranging and optical measurements together. Operator ranging means that the operator flying the satellite has determined its position by measuring the time it takes for light or radio waves to travel to it and bounce back.

Finally, Fig. 2.17 shows how optical and passive RF observations can be fused to further enhance the already accurate orbit solution that passive RF sensors can provide.

As can be inferred from Tab. 2.3 and Fig. 2.13 through Fig. 2.15, optical sensors are generally able to observe normal to the line of sight (generally in-track and cross-track positions) but not along the line of sight (≈ radial direction). This is why the depicted uncertainty volumes are primarily aligned with the radial direction. That said, the in-track positions observed by a series of optical tracking observations over successive orbits serve to refine the semi-major axis and eccentricity of the orbit, thereby allowing good radial solutions to be estimated as well.

Conversely, trackers that measure or can directly estimate relative range (radars, passive RF, Light Detection and Ranging (LIDAR), operator transponder ranging) predominantly observe the radial position ($\approx$ along the line of sight), leaving the larger uncertainties as being generally normal to the line of sight (both in-track and cross-track).

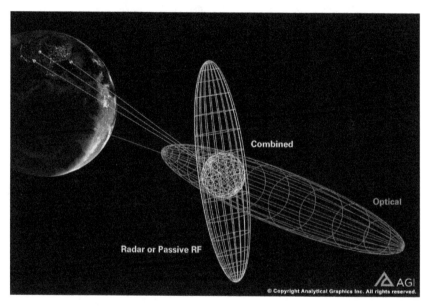

Figure 2.15: Benefit of fusing radar and optical observations together as opposed to a radar-only orbit solution or an optical-only solution

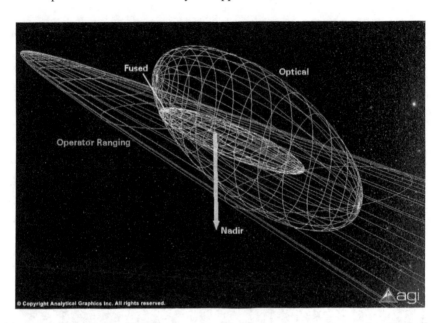

Figure 2.16: Benefit of fusing operator transponder ranging and optical observations together as opposed to an operator transponder ranging-only orbit solution or an optical-only solution

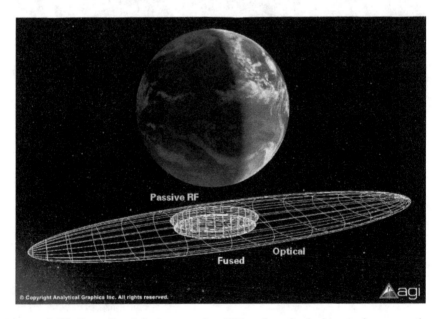

Figure 2.17: Benefit of fusing passive RF and optical observations together as opposed to a passive RF-only orbit solution or an optical-only solution

2.16 THE CASE FOR A SENSE OF URGENCY FOR IMPROVED SSA AND STM

In today's complex space operations environment, the benefits we derive from space and the welfare of our astronauts, spacecraft, commercial space industry, and general public are vulnerable. Spacecraft operators are not receiving the decision-quality SSA and STM required to conduct safe operations and ensure space sustainability.

The status quo is no longer sufficient given current flight safety limitations and in light of new knowledge and anticipated increases in space traffic. Flight safety services produce too many false alarms and miss too many serious threats. False alarms require spacecraft operators to squander precious resources and fuel, even as 96% of potentially lethal space objects remain untracked. This is compounded by outdated space-tracking algorithms, insufficient quality control, and a lack of transparency that all degrade flight safety. We have allowed legacy tools and algorithms to dictate the quality of SSA that operators receive. It is high time that we agree upon a set of key requirements an STM system must meet, and then "make it so."

Yet it is understandable how we got into this situation; where SSA and STM are concerned, the legacy tracking and analysis systems were not built using a comprehensive, top-down, requirements-based approach. While the Space Surveillance Network (SSN) is a great contribution to long-term sustainability, it is important to realize that this system was largely constructed in a piecemeal fashion. Many of the SSN's dedicated, collateral, and contributing sensors were originally designed and operated for non-SSA or STM purposes. It is unrealistic to think that such a system could adequately address the many challenges of SSA and STM.

We will now examine how our current and anticipated flight safety needs are unmet and urgently require better SSA and STM systems.

2.16.1 The current space debris environment

As enthralling as the movie *Gravity* was, it doesn't provide a realistic background and foundation upon which to build a solid space debris policy. This Hollywood depiction of space debris situations - similar to other exaggerated depictions such as such as Fig. 2.18 - would lead one to believe that spacecraft collisions in space happen frequently and are inevitable. This is false and misleading.

At the same time, one aspect of the film *Gravity* does ring true: collisions can have a severe impact on the space environment; in-space collisions and explosions have contributed to a dramatic increase in the density of objects in space. The crowding in spatial density of publicly tracked objects (the number of space objects per unit volume) is depicted in Fig. 2.19. There are multitudes more that we cannot yet accurately track, and colliding with these objects can also terminate a spacecraft's mission. In a metaphorical sense, we have built a highway in space over the past 60 years, and it is slowly filling up with cars. The sky is not falling yet, but as the number of fragments in certain orbit regimes has increased by a factor of 100 in 10 short years [23], it soon could be.

Figure 2.18: Depiction overstating the space debris problem (source: Adobe)

Publicizing a risk can be harmful if it causes citizens to feel overwhelmed in the face of that risk. Can the state of space debris really be that bad? After all, we operate every day in space, and it's not as if space collisions are regularly in the public eye.

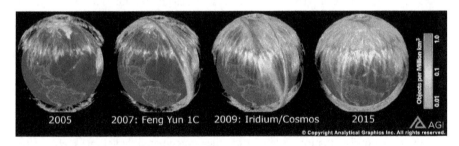

Figure 2.19: Comparison of STM attribute definitions by source

2.16.2 Collisions are already occurring

In reality, such collisions have occurred, and will continue to occur, in both LEO and GEO. Two of the most serious collisions were the intentional Chinese anti-satellite intercept of the Feng Yun spacecraft in 2007 and the accidental Iridium/Cosmos collision of 2009. The intense banding in the two centre images of Fig. 2.19 demonstrates debris from these collisions. Russia reports that in 2019, there were 63 violations of the 4-km safety radius established around the International Space Station [24]. Recurring close approaches between massive debris objects (e.g. spent upper stages and dead spacecraft) occur with regularity in space [25]. Although such events often go unreported, operators are increasingly observing spacecraft collisions with debris too small to officially track.

While such collisions are troubling and could terminate a spacecraft's mission, further increases in the population of spacecraft and debris fragments may at some point trigger the Kessler Syndrome, a chain reaction of collisions that would adversely impact our ability to operate in space by degrading GNSS, communications, earth imaging, and other critical services. In this scenario, the collision of two massive debris pieces generate so many large, heavy debris fragments that those fragments in turn collide with other substantial spacecraft or rocket bodies, and these secondary collisions lead to tertiary collisions and so forth, with each cascading generation of collisions yielding fragments of such sufficient mass and quantity that the chain reaction is self-sustaining.

Such a cycle is referred to as an "ecological threshold," a point at which a relatively small change or disturbance in external conditions can cause a rapid, exponential change in an ecosystem. In this case, that ecosystem is space. Once the ecological threshold is surpassed, the ecosystem may no longer be able to return to its previous state by means of its inherent resilience (e.g., in space, the ability of drag, gravitational resonances, solar radiation pressure, and other natural forces to "self-cleanse" and remove debris from orbit). Under these conditions, the Earth's orbital environment would be irreparably harmed.

While it is unclear whether we have ventured past a tipping point for safe and efficient space operations in any orbital regime, two things have become very clear. First, there are continual and substantive collision risks in both LEO and GEO orbit regimes. Second, mitigation of this risk requires satellite operators, space object tracking entities, and flight dynamicists to be ever vigilant and expend considerable resources and attention to ensure safe and efficient use of space for current and future generations.

2.16.3   The far-reaching effects of collisions and explosions in space

A common public and space policymaker/operator misconception is that spacecraft fragmentations tend to be localized events that only affect spacecraft in similar orbit altitudes or, in the case of GEO, similar geographic regions. More recently, it has become clear just how far-reaching a space collision or explosion can be, with collision and explosion fragmentation events adversely affecting the operability and commercial viability in space across all orbital regimes and GEO longitudinal locations [26, 27].

2.16.3.1   Two real-world LEO collisions and a hypothetical GEO case study

Let's examine how collision-induced fragmentation events can adversely affect the space population. In the following portrayals, volumetric regions denote portions of space that fragments may potentially occupy as a function of time. The likelihood that a fragment may be present at a specific position and time of interest is modelled using a bivariate Probability Density Function (PDF) of fragment characteristic length and

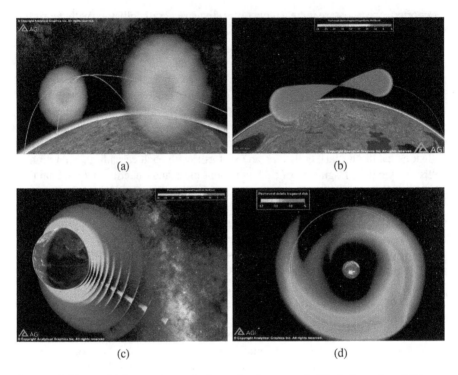

(a)

(b)

(c)

(d)

Figure 2.20: (a) Volumetric evolution of a fragment's likely location 170 seconds after the Chinese ASAT LEO kinetic intercept of Fengyun 1C (b) Volumetric aggregation of a fragment's likely location during the first 4 minutes after Fengyun 1C ASAT intercept (c) Volumetric aggregation of potential debris field during the first 3 hours after Fengyun 1C ASAT intercept (d) Volumetric aggregation of potential debris field during the first 26 hours after a hypothetical GEO collision

imparted velocity from the fragmentation event. The details of this calculation are not included for brevity but can be found in the aforementioned references [25, 26]. But it is worth underscoring that each integer decrement in the likelihood scale represents a tenfold decrease in likelihood of a fragment's presence. So, while these figures accurately depict the huge expanse of space that fragments can occupy, be mindful that the furthest reaches often represent a very low (but non-zero) likelihood of a fragment's presence.

The aforementioned Chinese Anti-Satellite (ASAT) intercept of Fengyun 1C is shown in Fig. 2.20a through Fig. 2.20c, a hypothetical GEO collision is shown in Fig. 2.20d, and the Iridium 33/Cosmos 2251 collision is shown in Fig. 2.21. Note that some of these figures are based upon different logarithmic gradient scales to best portray the volume at the time in question.

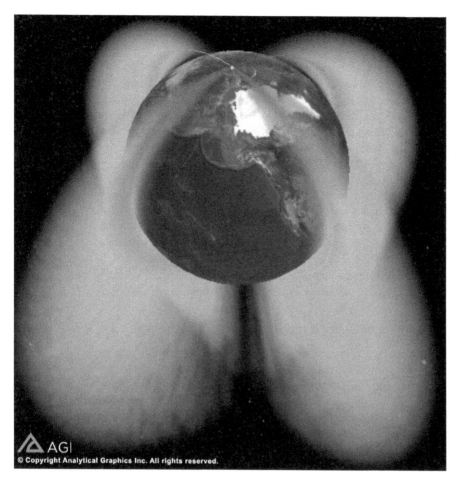

Figure 2.21: Volumetric aggregation of a fragment's likely location during the first three hours after the Iridium 33/Cosmos 2251 collision

Perhaps most noteworthy in all these depictions is that debris from LEO and GEO collisions can potentially affect all orbital regimes. The two real-world collisions have the potential to send debris fragments as high as 26,000 km in altitude, while the GEO collision can send debris fragments down to the Earth's surface and envelop much of the GEO belt within a day after the event.

While the sub-volume corresponding to the LEO post-collision high-altitude dispersal (Fig. 2.20c and Fig. 2.21) indicates a lower likelihood of its presence there, these portrayals nevertheless clearly indicate that both LEO and GEO collisions and explosions pose a global threat to the long-term sustainability of space activities.

2.16.4 Operators are already struggling to operate responsively and identify and avoid collisions

Past collisions of operational spacecraft and the extremely close approach of two dead spacecraft on 29 January 2020 are proof that today's approach to safety-of-flight is not enough. Even though avoidance of debris-generating collisions is a central pillar of the long-term sustainability of space activities, today's LEO and GEO operators frequently cannot tell when collision avoidance manoeuvres are required, often due to limitations in orbital accuracy, precision, completeness, timeliness, and transparency in both operator and State-provided data.

In addition, an estimated 40% of all spacecraft and 35% of all upper stages are not disposed of upon completion of their mission [28] to limit their post-mission presence in the Low Earth Orbit (LEO) and GEO protected regions. Yet we now recognize the probability of successful post-mission disposal of spacecraft to be one of the most critical parameters to ensure space sustainability, and disposal rates as high as 95% may be required to ensure long-term sustainability of space activities. And at times, spacecraft and upper stages are not properly passivated to deplete all energy sources that could lead to explosive fragmentation events.

2.16.5 Potential for a tenfold increase in active satellites

We are entering a phase of unparalleled change. Business as usual is no longer an option as the New Space era rapidly dawns. Many are unaware that there is a bow wave of upcoming large constellations. Plans have been filed with the International Telecommunication Union (ITU) and the Federal Communications Commission (FCC) or announced in the media to build, launch and operate over 58,000 spacecraft within the next ten years alone, a tenfold increase in the number of operational spacecraft (Fig. 2.22). We acknowledge that only a portion of these spacecraft applications will be realized as operational spacecraft, but even if only 10 to 50% of these constellations become operational, we could easily see an active spacecraft population in the next decade that is between four and ten times larger than is flying today. This year alone, the active space population is on track to nearly double. Of these 58,000 spacecraft applications, U.S. companies have proposed 66 times more than any other country, which equates to 25 times more spacecraft than all active spacecraft flying today.

This is an exciting time for space, but it demands that we get prepared on the regulatory, SSA, and STM fronts. The technically advanced spacecraft flying in these large constellations will also operate using more sophisticated technology. Low-thrust propulsion will be the rule rather than the exception, and autonomous on-board navigation and collision-avoidance systems will be standard. Low-thrust manoeuvres can cause difficulties for older SSA systems with no manoeuvre estimation.

Large constellations will experience millions of close approaches, requiring thousands of avoidance manoeuvres, with many being as close or closer than the 29 January close approach of two dead spacecraft, the Infra-Red Astronomy Satellite and the Gravity Gradient Stabilization Experiment 4 spacecraft. Our updated research results [29] shown in Fig. 2.23 portray the anticipated high rates of collisions, 3 km

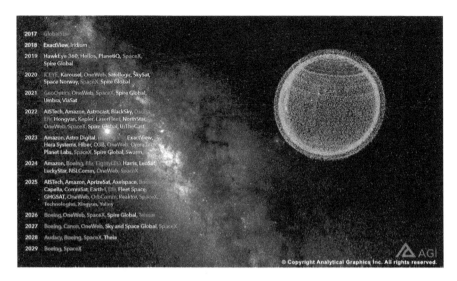

Figure 2.22: Large constellations for which applications have been filed, 2019 – 2029

warnings, and 1 km manoeuvres required for large constellations against the currently tracked catalogue (middle 3 columns) and estimated catalogue above 1 cm (right 3 columns). Left unchecked, many collisions are likely to occur. For example, it has been estimated that the developing Starlink constellation of 4,425 spacecraft will experience two million close approaches over a ten-year mission, resulting in six potentially environment-altering collisions with currently tracked debris if left unmitigated, and an additional 71 potentially mission-terminating collisions against the full population down to 1 cm in size.

While the global population of active spacecraft will grow over the next decade, we do have a few years to prepare for this upcoming rapid growth. But we must take steps now.

2.16.6 Potential for a tenfold increase in tracked debris

Only an estimated 4% of the LEO and GEO space populations are currently being tracked. Outdated space-tracking algorithms along with insufficient quality control and service level availability further degrade the space catalogue's completeness, accuracy, timeliness, and transparency.

These deficiencies may soon be addressed through the impending realization of the operational Space Fence and promising advances made by commercial radar tracking companies. This means that the number of tracked space objects could soon increase tenfold if we only consider objects that are already in space but have heretofore been untrackable.

Operator	# S/C	Alt (km)	Current (c10 cm) RSO catalog average number			~200,000 (≈2 cm) RSO catalog average number		
			Estimated collisions in 10 years	3km warnings in 10 years	1km maneuvers in 10 years	Estimated collisions in 10 years	3km warnings in 10 years	1km maneuvers in 10 years
AISTech_Danu	300	591	0.07	479,649	53,294	0.19	4,635,985	515,109
Amazon	3,236	590	0.18	3,768,872	418,764	0.09	36,120,810	4,013,423
Boeing_1	1,120	1,200	0.14	331,965	36,885	1.09	4,739,224	526,580
Boeing_2	1,210	550	0.10	234,358	26,040	0.84	3,646,359	405,151
Boeing_3	1,000	585	0.23	1,812,814	201,424	0.59	16,903,756	1,878,195
CommSat	800	600	0.07	1,362,606	151,401	0.03	12,835,938	1,426,215
ExactView	72	820	0.21	326,914	36,324	1.10	2,768,355	307,595
Hongyan	300	1,100	0.04	241,520	26,836	0.16	3,434,841	381,649
Iridium	85	781	0.06	399,037	44,337	0.12	2,514,772	279,419
LuckyStar	156	1,000	0.02	318,736	35,415	0.01	2,616,385	290,710
OneWeb	2,560	1,200	0.32	754,868	83,874	2.49	10,832,864	1,203,652
OneWeb_next	720	1,200	0.17	286,598	31,844	1.69	4,726,261	525,140
Satellogic	300	477	0.02	236,040	26,227	0.02	2,254,977	250,553
SpaceX	4,425	1,200	6.43	2,050,452	227,828	77.73	30,310,084	3,367,787
SpaceX_VLEO	1584	550	3.45	1,101,453	122,384	35.63	13,894,159	1,543,795
Space_X_M-T	20,940	500	43.13	13,783,896	1,528,211	404.53	157,747,388	17,527,488
Space_X_U-W	9,000	330	0.93	347,030	38,559	21.86	10,053,221	1,117,025
Theia	211	775	1.08	783,728	87,081	7.57	7,520,310	835,590
Xingyun	156	1,000	0.04	360,898	40,100	0.06	2,831,654	314,628
Yaliny	140	1,000	0.03	321,780	35,753	0.05	2,599,648	288,850

Figure 2.23: Collision, warning, and manoeuvre rates for the top 20 proposed large constellations

2.16.7 Emergence of Rendezvous and Proximity Operations and On-Orbit Servicing

The emergence of Rendezvous and Proximity Operations (RPO) and On-orbit Servicing (OOS) spacecraft adds a further layer of complexity. The promising commercial flight of the on-orbit servicer Mission Extension Vehicle (MEV-1) and other Active Debris Removal (ADR) platforms preparing for flight, further underscores our increasingly complex future space environment.

2.16.8 More commercial and international space operations centers

Some estimate that the global SSA market could reach $1.1B by 2025. Commercial American SSA and STM service providers are on the leading edge of this global market, applying innovative, cost-saving hardware, algorithms, and software to these domains. As a direct result of these innovations, space catalogues are growing with the inclusion of smaller debris with orbits known with greater accuracy than commercial spacecraft operators have ever achieved before. Unfortunately for U.S. commercial SSA providers, their government has not succeeded in finding ways to incorporate such commercial SSA services into government safety-of-flight analyses and products. Providing U.S. government SSA and STM services at no cost to spacecraft operators, while promoting flight safety for the benefit of all, represents direct competition with American companies that commercially provide SSA and STM services (who may go out of business soon if this competition is not addressed).

2.16.9 Greater need to coordinate space traffic than ever before

Collectively, this explosive growth in the number of spacecraft will also change the statistics of the types of collisions, increasing the number of active-on-active spacecraft conjunctions to an all-time high. This further highlights the need for robust, protected, and verifiable information pooling and standardization.

2.16.10 Increase in the number of space actors

We are also in the midst of an explosion in the number of space actors. The popularity of CubeSats and mass-produced small satellites is lowering the costs of procuring and launching spacecraft.

2.16.11 Increasing spacecraft and operating complexities

The high conjunction rates anticipated for large constellations will naturally fuel the desire for as-yet-unproven automated collision avoidance. Automated avoidance would mean that a spacecraft could decide on its own what optimal avoidance manoeuvre to conduct and when. However, if this is not shared with the other spacecraft operator, then the two spacecraft could potentially both steer directly into each other.

There are advances in spacecraft propulsion. Large constellations will use low-thrust propulsion as the rule rather than exception. Besides requiring more avoidance time, low-thrust manoeuvres can cause difficulties for older SSA systems that don't possess manoeuvre estimation capabilities.

Many CubeSats manoeuvre by differential drag and drag augmentation sail approaches. In differential drag, the operator changes spacecraft attitude relative to other satellites in their fleet to manoeuvre by "catching the wind." Drag augmentation sails deploy to greatly increase drag, causing the spacecraft to reenter quicker than it otherwise would. Both of these techniques can challenge some SSA systems.

2.16.12 More advanced SSA processing algorithms and scalable architectures

Despite having been established for centuries, much progress continues to be made in the development of advanced SSA-related astrodynamics, orbit determination, and collision risk assessment algorithms. The application of sequential filters with built-in manoeuvre detection and characterization allow SSA systems to be much more responsive to the constantly manoeuvring active space population. Those advanced algorithms need to be implemented in scalable architectures in the burgeoning space population associated with the New Space era.

2.17 SUMMARY OF SSA AND STM

We discussed the variety of definitions associated with the terms "SSA" and "STM", described their major components, and illustrated why SSA and STM are challenging tasks. We explained the standard conjunction assessment process and enumerated the commercial, national, and international entities who provide such services. We then turned our attention to the importance of space standards, commercial best practices, the criticality of global space data exchange, and the need for advanced algorithms and data fusion. Finally, we characterized the current space environment and the current and future flight safety challenges we face.

Global access to timely, accurate, comprehensive, transparent, highly-available and standards-based SSA and STM services will be foundational to ensuring safety-of-flight, mitigating RFI and achieving long-term sustainability of space activities, given the anticipated explosive growth in both the number of active spacecraft and our knowledge of the debris population within the decade.

While the situation is not yet dire if properly managed, substantive and ongoing collision risks exist in both LEO and GEO. These collision risks and the far-reaching and lasting effects of fragmentation events are sobering. While we are globally challenged by SSA today, extensive leveraging of advanced algorithms, research, and crowd-sourcing and data fusion of spacecraft operator and government and commercial SSA data can provide us with the critical capabilities and data required to address these SSA challenges.

ACRONYM

18SpCS United States Air Force 18th Space Control Squadron

AoR Area of Responsibility

ASAT Anti-Satellite weapon

ATC Air Traffic Control

CA Conjunction Assessment

COLA Collision Avoidance

CCSDS Consultative Committee for Space Data Systems

CNES Centre national d'études spatiales

COPUOS Committee for the Peaceful Use of Outer Space

CSpOC United States Combined Space Operations Center

DoD United States Department of Defense

ESA European Space Agency

ETM Upper E Traffic Management (also High E, High Altitude or Near-Space Traffic Management)

EU European Union

EUSST European Union Space Surveillance and Tracking

FCC United States Federal Communications Commission

FHR Fraunhofer Institute for High Frequency Physics and Radar Techniques

GEO Geostationary Earth Orbit

GNSS Global Navigation Satellite System

GRAVES Grand Réseau Adapté à la Veille Spatiale

IADC Inter-Agency Space Debris Coordination Committee

ICAO International Civil Aviation Organization

ISO International Organization for Standardization

ISON International Scientific Observation Network

ITU International Telecommunication Union

LCOLA Launch Collision Avoidance

LEO Low Earth Orbit, typically defined as below 2000 km altitude above spherical Earth

LIDAR Light Detection and Ranging

LTS Long-Term Sustainability of space activities

MSL Mean Sea Level

NASA National Aeronautics and Space Administration

OA Observation Association

OD Orbit Determination

OODA Observe/Orient/Decide/Act decision processing loop

OOS On-Orbit Servicing

RCS Radar Cross Section

RF Radio Frequency

RFI Radio Frequency Interference

RPO Rendezvous and Proximity Operations

RSO Resident Space Object

SDA Space Data Association (context specific)

SDA Space Domain Awareness (context specific)

SDC Space Data Center

SEM Space Environment Management

SGP Simplified General Perturbations orbit theory (and associated propagator)

SOCRATES Satellite Orbital Conjunction Reports Assessing Threatening Encounters in Space

SPD United States Space Policy Directive

SSA Space Situational Awareness

SSC Space Safety Coalition

SSN Space Surveillance Network

SSR Space Sustainability Rating

SST Space Surveillance and Tracking

SSO Satellite Servicing Operations

STC Space Traffic Coordination

STM Space Traffic Management

TAROT Télescopes à Action Rapide pour les Objets Transitoires

TIRA Tracking and Imaging Radar

UN United Nations

USAF United States Air Force

USSTRATCOM United States Strategic Command

UTM Unmanned Aircraft System Traffic Management

Vmag Visual Magnitude

WNCO Willing Non-Contributor Operator

WEF World Economic Forum

GLOSSARY

18th Space Control Squadron (18 SPCS): A space control unit located at Vandenberg AFB, California, tasked with providing 24/7 support to the space sensor network (SSN), maintaining the space catalogue and managing USSPACECOM's space situational awareness (SSA) sharing program to United States, foreign government, and commercial entities.

Active Debris Removal (ADR): A process in which spacecraft are deployed to capture and de-orbit larger pieces of debris and out-of-service satellites.

Anti-satellite weapon (ASAT): A space weapon designed to incapacitate or destroy satellites for strategic military purposes.

Area of Responsibility (AOR): A pre-defined geographic region assigned to combatant commanders of the Unified Command Plan (UCP), that are used to define an area with specific geographic boundaries where they have the authority to plan and conduct operations; for which a force, or component commander bears a certain responsibility.

CelesTrak: The first public space data portal; founded in 1985 by Dr. T.S. Kelso.

Combined Space Operations Center (CSpOC): A U.S.–led multinational space operations centre that provides command and control of space forces for United States Space Command's Combined Force Space Component Command.

Commercial Space Operations Center (ComSpOC): A space situational awareness (SSA) facility developed by Analytical Graphics, Inc. (AGI), that tracks space objects to monitor threats and sustain safety in space.

Collision: An incident occurring when two objects of any type impact each other on orbit.

Collision Avoidance (COLA): Procedures undertaken, such as a manoeuvre of an active payload, to mitigate the risk of an impending collision in a conjunction event.

Conjunction: A close encounter in orbit between two resident space objects.

Conjunction Assessment (CA): Spaceflight safety analyses that identify a close encounter between space objects.

Conjunction Data Message (CDM): Internationally-standardized mechanism for data exchange of conjunction information between originators of conjunction assessment data and satellite owner/operators; the means of notification of a potential collision risk.

Consultative Committee for Space Data Systems (CCSDS): A multi-national forum for the development of communications and data systems standards for spaceflight; founded in 1982.

European Union Satellite Centre (EU SatCen): EU centre that provides fast and reliable analysis of satellite data in order to face current security challenges.

European Union Space Surveillance & Tracking (EUSST): The EU's capacity to detect, catalogue and predict the movements of space objects orbiting the Earth.

Geostationary Earth Orbit (GEO): A special case of geosynchronous orbit with a circular geosynchronous orbit in Earth's equatorial plane and an orbital period that matches Earth's rotation on its axis, 23 hours, 56 minutes, and 4 seconds (one sidereal day).

Global Navigation Satellite System (GNSS): A constellation of satellites that provide geo-spatial positioning to many devices autonomously, allowing electronic devices with the appropriate receivers to determine their precise location on the surface of the Earth.

Inter-Agency Space Debris Coordination Committee (IADC): An inter- governmental forum whose aim is to co-ordinate efforts to deal with debris in orbit around the Earth; founded in 1993.

International Civil Aviation Organization (ICAO): A United Nations (UN) specialized agency managing the administration and governance of the Convention on International Civil Aviation (Chicago Convention); established by States in 1944.

International Scientific Optical Network (ISON): An international project, currently consisting of about 30 telescopes at about 20 observatories in about ten countries which have organized to detect, monitor and track objects in space.

International Organization for Standardization (ISO): An international standard-setting body composed of representatives from various national standards organizations, which promotes worldwide proprietary, industrial, and commercial standards; founded in 1947.

Long-Term Sustainability of Outer Space Activities (LTSSA): A set of guidelines (adopted by the United Nations Committee on the Peaceful Uses of Outer Space) concerning the policy and regulatory framework for space activities; safety of space operations; international cooperation, capacity-building and awareness; and scientific and technical research and development.

Low Earth Orbit (LEO): An Earth-centred orbit with an altitude of 2,000 km (1,200 mi) or less (approximately one-third of the radius of Earth), or with at least 11.25 periods per day (an orbital period of 128 minutes or less) and an eccentricity less than 0.25.

Medium Earth orbit (MEO): The region of space around Earth above low Earth orbit (altitude of 2,000 km (1,243 mi) above sea level) and below geosynchronous orbit (altitude of 35,786 km (22,236 mi) above sea level); sometimes called intermediate circular orbit (ICO).

Moon Treaty: Agreement Governing the Activities of States on the Moon and Other Celestial Bodies (1979); failed to be ratified by any major space-faring nation.

New Space: A movement and philosophy encompassing a globally emerging private spaceflight industry; specifically, the term is used to refer to a global sector of new aerospace companies and ventures working independently of governments and traditional major contractors to develop faster, better, and cheaper access to space and spaceflight technologies, driven by commercial, as distinct from political or other, motivations to broader, more socioeconomically-oriented, ends.

Observation Association (OA): The process of determining which sensor observations/tracks belong to which objects and should be used to update the orbit determination of the objects.

On-Orbit satellite Servicing (OOS): The repair, refurbishment, refuelling, upgrading and/or assembly of satellites (or satellite components) which have already been launched.

Orbit Data Message (ODM): Internationally-standardized mechanism for data exchange of a space object's orbit, covariance, physical characteristics, manoeuvres, and state transition matrices.

Orbit Determination (OD): The estimation of orbits of objects such as moons, planets, and spacecraft.

Outer Space Treaty: A treaty that forms the basis of international space law; formally the Treaty on Principles Governing the Activities of States in the Exploration and Use of Outer Space, including the Moon and Other Celestial Bodies.

Radio Frequency Interference (RFI): Radio frequency interference is the conduction or radiation of radio frequency energy that causes an electronic or electrical device to produce noise that typically interferes with the function of an adjacent device; it also refers to the disruption of the normal functionality of a satellite due to the interference of radio astronomy.

Registration Convention: Convention on Registration of Objects Launched into Outer Space (1976).

Rendezvous and Proximity Operations (RPO): An orbital manoeuvre during which two spacecraft arrive at the same orbit and approach to a very close distance (e.g. within visual contact); requires a precise match of the orbital velocities and position vectors of the two spacecraft, allowing them to remain at a constant distance through orbital station-keeping; may or may not be followed by docking or berthing, procedures which bring the spacecraft into physical contact and create a link between them.

Rescue Agreement: Agreement on the Rescue of Astronauts, the Return of Astronauts and the Return of Objects Launched into Outer Space (1968).

Resident Space Object (RSO): An on-orbit satellite of any type – satellites (active or inactive), spent rocket-bodies, or fragmentation debris.

Satellite Catalogue: A "1 to N" numbering and identification system of on-orbit objects.

Satellite Orbital Conjunction Reports Assessing Threatening Encounters in Space: SOCRATES, an online conjunction assessment tool; added to CelesTrak in 2004.

Space Data Association (SDA): A commercial spacecraft operator self-formed international organization to support the controlled, reliable and efficient pooling of space operator and SSA data and information critical to the safety and integrity of the space environment using a Data Lake approach in a computationally and legally protected framework.

Space Data Center (SDC): The SDA platform, operated by SDA's trusted technology partner, AGI, that ingests flight dynamics information from the member companies as well as other available sources of space object information to provide conjunction assessment and warning services.

Space Domain Awareness (SDA): The maintenance of a catalogue and associated orbit information along with the identification, characterization, and understanding of any factor or behaviour, passive or active, associated with the space domain that could affect space operations, thereby potentially impacting the nation's security, safety, economy, and environment.

Space Liability Convention: Convention on International Liability for Damage Caused by Space Objects (1972).

Space Policy Directive 3 (SPD-3): A space directive focused on STM, which shifts responsibility for providing space situational awareness (SSA) data to satellite operators from the Department of Defense (DoD) to the Department of Commerce (DoC); provides guidelines and direction to ensure that the United States is a leader in providing a safe and secure environment as commercial and civil space traffic increases; signed by President Trump in 2018.

Space Safety Coalition (SSC): An ad hoc coalition of companies, organizations, and other government and industry stakeholders that actively promotes responsible space safety through the adoption of relevant international standards, guidelines and practices, and the development of more stringent and effective voluntary space safety guidelines and best practices.

Space Situational Awareness (SSA): The knowledge and characterization of space objects and their operational environment to support safe, stable, and sustainable space activities.

Space Surveillance Network (SSN): A world-wide, globally dispersed network of radar, electro-optical and passive radio frequency (RF) sensors that detects, tracks, catalogues and identifies artificial objects orbiting Earth.

Space Sustainability Rating (SSR): A score representing a mission's sustainability as it relates to debris mitigation and alignment with international guidelines; an initiative of the World Economic Forum (WEF).

Space Traffic Management (STM): The planning, coordination, and on-orbit synchronization of activities to enhance the safety, stability, and sustainability of operations in the space environment.

Technology Readiness Level (TRL): A method for estimating the maturity of technologies during the acquisition phase of a program; developed at NASA during the 1970s.

Two-Line Element set (TLE): A data format encoding a list of orbital elements of an Earth-orbiting object for a given point in time, the epoch.

United Nations Committee on the Peaceful Uses of Outer Space (UN COPUOS): Set up by the General Assembly in 1959 to govern the exploration and use of space for the benefit of all humanity: for peace, security and development; tasked with reviewing international cooperation in peaceful uses of outer space, studying space-related activities that could be undertaken by the United Nations, encouraging space research programmes, and studying legal problems arising from the exploration of outer space.

United States Space Command (USSPACECOM): A unified combatant command of the United States Department of Defense, responsible for military operations in outer space, specifically all operations 100 kilometres and above mean sea level.

United States Space Force (USSF): The space warfare service branch of the United States Armed Forces that organizes, trains, and equips space forces in order to protect U.S. and allied interests in space and to provide space capabilities to the joint force.

FURTHER READING

European Space Policy Institute (2020). *ESPI Report 71 - Towards a European Approach to Space Traffic Management*. Editor and Publisher: ESPI. ISSN: 2218-0931 (print) and 2076-6688 (online).

The MITRE Corporation, the US Chamber of Commerce, and the National Cybersecurity Center (2020). *Understanding the Influence of Space Situational Awareness on Commercial Space Development*.
https://www.uschamber.com/report/understanding-the-influence -of-space-situational-awareness-commercial-space-development.

Undseth, M., C. Jolly and M. Olivari (2020). *Space Sustainability: The Economics of Space Debris in Perspective*. OECD Science, Technology and Industry Policy Papers, No. 87, OECD Publishing, Paris, https://doi.org/10.1787/a339de43-en.

European Space Agency (2019). *ESA's Annual Space Environment Report* https://www.sdo.esoc.esa.int/environment_report/Space_Environment_Report_latest.pdf. Prepared by ESA Space Debris Office: Reference: GEN-DB-LOG-00271-OPS-SD.

European Space Agency (2008). *Europe's Eyes on the Skies – The Proposal for a European Space Surveillance System*. ESA Publications, Bulletin, No. 133, http://www.esa.int/esapub/bulletin/bulletin133/bul133f_klinkrad.pdf.

Oltrogge, D.L. and Alfano, S. (2019). *The technical challenges to better Space Situational Awareness and Space Traffic Management*. Journal of Space Safety Engineering https://doi.org/10.1016/j.jsse.2019.05.004

Oltrogge, D.L. (2020). *Space Situational Awareness: Key Issues in An Evolving Landscape*. U.S. House Subcommittee on Space and Aeronautics testimony on 11 February 2020.

Oltrogge, D.L., (2019). *Addressing Space Traffic Management at Multinational, National and Industry Levels.* World Space Forum, Vienna.

Oltrogge, D.L. and Cooper, J.A. (2018). *Practical Considerations and a Realistic Framework for a Space Traffic Management System.* 18th Australian Aerospace Congress, Melbourne, Australia.

Kelso, T.S., and Oltrogge, D.L. (2018). *The Need for Comparative SSA.* IAC-18-A6.7.8, International Astronautical Congress (IAC), Bremen, Germany.

Oltrogge, D.L. (2018). *The "We" Approach to Space Traffic Management.* The 15th International Conference on Space Operations, Marseilles, France.

Berry, D.S., and Oltrogge, D.L. (2018). *The Evolution of the CCSDS Orbit Data Messages.* The 15th International Conference on Space Operations, Marseilles, France.

Oltrogge, D.L., Alfano, S., Law, C. , Cacioni, A., Kelso, T.S. (2018). *A comprehensive assessment of collision likelihood in Geosynchronous Earth Orbit.* Acta Astronautica (2018), doi:10.1016/j.actaastro.2018.03.017.

Alfano, S. and Oltrogge, D. (2018). *Probability of collision: Valuation, variability, visualization, and validity.* Acta Astronautica (2018), doi:10.1016/j.actaastro.2018.04.023.

Reesman R., Gleason M.P., Bryant, L., Stover, C. (2020). *Slash the Trash: Incentivizing Deorbit.* Center for Space Policy and Strategy. https://aerospace.org/sites/default/files/2020-04/Reesman_SlashTheTrash_20200422.pdf

Bibliography

[1] National Space Policy of the United States of America, 28 June 2010. https://www.space.commerce.gov/policy/national-space-policy/.

[2] O. Stelmakh-Drescher. Space Situational Awareness and Space Traffic Management: Towards Their Comprehensive Paradigm. In *Space Traffic Management Conference, Embry-Riddle Aeronautical University, 17 November*, 11 2016.

[3] European Space Agency. "SSA Programme Overview", ESA Space Debris website (accessed on April 2, 2020). https://www.esa.int/Safety_Security/SSA_Programme_overview.

[4] C. Bonnal, L. Francillout, M. Moury, U. Aniakou, J. D. Perez, J. Mariez, and S. Michel. CNES technical considerations on space traffic management. *Acta Astronautica*, 167:296–301, 2020. https://doi.org/10.1016/j.actaastro.2019.11.023.

[5] "SSA - SatCen - European Union", SatCen website (accessed on April 2, 2020). https://www.satcen.europa.eu/page/ssa.

[6] Space Foundation website (accessed on April 2, 2020). `https://www.spacefoundation.org/space_brief/space-situational-awareness/`.

[7] United States Space Policy Directive 3. National Space Traffic Management Policy, 18 June 2018. `https://www.whitehouse.gov/presidential-actions/space-policy-directive-3-national-space-traffic-management-policy`.

[8] K.U. Schrogl, C. Jorgenson, J. Robinson, and Soucek A. "The IAA Cosmic Study on Space Traffic Management", United Nations Committee on the Peaceful Uses of Outer Space (UN COPUOS), 6 April 2017, Online. `http://www.unoosa.org/documents/pdf/copuos/lsc/2017/tech-10.pdf`.

[9] O. Stelmakh-Drescher. Space Situational Awareness and Space Traffic Management: Towards Their Comprehensive Paradigm. In *Space Traffic Management Conference, Embry-Riddle Aeronautical University, 17 November*, 11 2016.

[10] European Space Policy Institute. "ESPI Report 71: Towards a European Approach to Space Traffic Management", ISSN: 2218-0931 (print), 2076-6688 (online), January 2020.

[11] Marietta Benkö, Kai-Uwe Schrogl, Denise Digrell, and Esther Jolley. *Space Law: Current Problems and Perspectives for Future Regulation*, volume 2. Eleven International Publishing, 2005.

[12] Darren McKnight. Examination of spacecraft anomalies provides insight into complex space environment. *Acta Astronautica*, 10.036, 2017.

[13] The Space Safety Coalition. "best Practices for the Sustainability of Space Operations". `spacesafety.org`.

[14] D. Oltrogge. Space Data Association and SDA Conjunction Assessment Services. In *21st Improving Space Operations Support Workshop, Pasadena CA*, 2015.

[15] M. C. Smitham. The Need for a Global Space-Traffic-Control Service: An Opportunity for US Leadership. Technical report, Air War College Air University Maxwell AFB United States, Maxwell Paper No. 57, 2010. `http://www.au.af.mil/au/awc/awcgate/maxwell/mp57.pdf`.

[16] Julian P. McCafferty. "Development of a Modularized Software Architecture to Enhance SSA with COTS Telescopes". `https://scholar.afit.edu/cgi/viewcontent.cgi?article=1438&context=etdhttps://scholar.afit.edu/cgi/viewcontent.cgi?article=1438&context=etd`, 2016.

[17] "12 Space Warning Squadron AN/FPS-120, Two-sided, Solid-state, Phased-array Radar System (SSPARS) at Thule Air Base". Public Domain. `https://en.wikipedia.org/wiki/File:Thule_12th_sws.jpg`, 2020.

[18] "The 143.050mhz Graves Radar: a VHF Beacon for Amateur Radio". `https://fas.org/spp/military/program/track/graves.pdf`, 2020.

[19] "Télescopes à Action Rapide pour les Objets Transitoires. Credit: Par ESO — http://www.eso.org/public/france/images/tarot_lasilla201201b/, CC BY 4.0, https://commons.wikimedia.org/w/index.php?curid=48472093. https://fr.wikipedia.org/wiki/T%C3%A9lescope_%C3%A0_action_rapide_pour_les_objets_transitoires, 2020.

[20] "Gelände der Fraunhofer-Institute in Wachtberg", image licensed under the creative commons attribution 3.0 unported license, https://en.wikipedia.org/wiki/en:Creative_Commons. https://commons.wikimedia.org/wiki/File:Fraunhofer-Gel%C3%A4nde_Wachtberg.jpg, 2020.

[21] Chilbolton Observatory, Credit: Original uploader was drichards2 at English Wikipedia. Transferred from en.wikipedia to Commons by Mike Peel using CommonsHelper., CC BY-SA 3.0, https://commons.wikimedia.org/w/index.php?curid=17571830. https://en.wikipedia.org/wiki/Chilbolton_Observatory#/media/File:Chilbolton_Observatory_3GHz_Radar_Antenna.jpg, 2020.

[22] D. L. Oltrogge and J. A. Cooper. Practical Considerations and a Realistic Framework for a Space Traffic Management System. In *AIAC18: 18th Australian International Aerospace Congress , Australia*. Engineers Australia, Royal Aeronautical Society., 2018.

[23] D. Oltrogge and S. Alfano. Collision Risk in Low Earth Orbit. In *IAC-16, A6,2,1,x32763, 67th International Astronautical Congress, Guadalajara, Mexico*, 2016.

[24] "Russian Federation Space Debris Mitigation Activities", 57th Session of the UN Scientific and Technical Subcommittee, Vienna Austria. https://www.unoosa.org/documents/pdf/copuos/stsc/2020/tech-36E.pdf, 2020.

[25] D. McKnight, D.L. Oltrogge, S. Alfano, R. Shepperd, S. Speaks, and J. Macdonald. "the Cost of Not Doing Debris Remediation". In *IAC-19,A6,2,7,x48725, 70th International Astronautical Congress, Washington, D.C*, 2019.

[26] D.L. Oltrogge and D.A. Vallado. Application Of New Debris Risk Evolution And Dissipation (DREAD) Tool To Characterize Post-Fragmentation Risk. In *2017 Astrodynamics Specialist Conference, Stevenson WA, AAS*, pages 17–600, 22 August 2017.

[27] D.L. Oltrogge and D.A. Vallado. Debris Risk Evolution And Dispersal (DREAD) for Post-Fragmentation Modeling. In *Hypervelocity Impact Symposium, Destin, FL, USA*, 14–19 April 2019.

[28] The European Space Agency Space Debris Office. ESA's Annual Space Environment Report. in v.3.2, Issued 17 July 2019.

[29] S. Alfano, D.L. Oltrogge, and R. Shepperd. LEO Constellation Encounter and Collision Rate Estimation: An Update. In *2nd IAA Conference on Space Situational Awareness, IAA-ICSSA-20-0021*, 14 January 2020.

III

Dealing with Space Debris. Sociotechnical Concerns

Space Debris Sustainability: Understanding and Engaging Outer Space Environments

Michael Clormann

Technical University of Munich, Munich, Germany

Nina Klimburg-Witjes

Vienna University, Vienna, Austria

CONTENTS

IN recent years, space debris has become a matter of considerable importance within the public perception of spaceflight activities. As media coverage, outreach activities and stakeholder interest in orbital waste gain more and more relevance, so does the question of how space debris can be understood as a sociotechnical challenge that contemporary and future societies depending on space-based services need to address. Similar to problems like climate change or marine pollution, space debris appears as a sustainability issue of global magnitude that requires us to think about outer space in terms of a sustainable societal future. Yet, space debris also differs from such seemingly comparable challenges in some regards, as, for example, it is only to be understood within the context of recent space sector developments like the rise of New Space. Paying attention to security concerns as well as the specific ecological status of outer space environments, we outline possible avenues to painting a better picture of space debris' role in contemporary public and policy debates. Bringing to the table a perspective influenced by Science and Technology Studies (STS), we thereby highlight space debris to be a bidirectional risk phenomenon. We conclude, that broader societal engagement in facing the challenge of space debris might be vital for handling it efficiently and effectively and propose potential vectors for stakeholder participation.

3.1   INTRODUCTION

Space has become a place where sustainability is increasingly negotiated as an issue of security, as billions of people around the world rely on space systems to facilitate their daily life, from navigation to environmental services, from science to communication, crisis response, banking, and transport. However, the recent promises of a New Space orbital gold-rush are clouded by the legacy of "Old Space": The remains of decades of spaceflight activities have left an ever-growing pile of junk in the Earth's orbits – rocket components, defunct satellites, and propellant residues are just a few of them. These debris fragments can lead to the congestion of current and – even more important – future uses of outer space. Interfering with communication and navigation systems in orbit, they pose severe risks to the infrastructures modern societies rely on. The worst-case scenario concerning space debris predicts a future, where the orbits are becoming permanently impenetrable [1] to astronomical observation as well as any form of space activity aiming to leave the planet or utilize its orbits.

In this chapter, we will highlight two different aspects of space debris as a sociotechnical concern.

First, the current framing of space debris as a *sustainability* issue by space policy makers and in public debates. Here, concepts of sustainable technologies carry different and profound meanings in societal debates on the future of certain technologies or even industries as a whole, while in the early decades of the Space Age, sustainability was not a major issue, as launching objects into space and operating them on vast cosmic scales seemed enough of a challenge [2]. Today, the scenario of a gridlocked outer space due to space debris is beginning to receive attention not only in policy discourse, but in public perception as well.

As a "remnant" of the Space Age and its strategic political concerns, *security* today proves to be one of the focal points for articulating sustainability in outer space and especially towards space debris. Therefore, we suggest to, secondly, understand space debris as a twofold source of societal risk perception. Unlike other environmental sustainability challenges, space debris features two significant and opposite "risk vectors". On the one hand, the risk of damaged property or health through debris re-entry events is prominent in the public perception of space debris as a problem. On the other hand, debris is perceived as problematic as it not always re-enters within reasonable time-frames – posing for example a continuous risk to satellite constellations in orbit or optically polluting the night sky for astronomers or amateur stargazers.

In that sense, space debris might directly impede our individual safety living on the planet's surface while also threatening the orbital satellite infrastructures modern technologized societies need to function. The first risk vector is a 'downward' one: the source of risk being located in LEO environments to affect the planet below. The second risk vector can be seen as an 'upward' one: producing increasing risk levels within orbital environments through partially unregulated launches activities and ever increasing numbers of satellites that at some point decay into space debris and remain in orbit, potentially hampering current and future spaceflight activities and the essential services they provide.

In our conclusion, we will outline the implications of these risk configurations when it comes to public participation as well as possible ways of shaping a more societally inclusive way to address the challenge of space debris. We end by highlighting participation of stakeholders as a chance to achieve necessary sociotechnical resilience in sustaining outer space environments.

3.2 PERSPECTIVES FROM SCIENCE AND TECHNOLOGY STUD-IES

Science and Technology Studies (STS) are a relatively young field in the spectrum of social sciences, at the intersections of sociology, political science, anthropology, history and the philosophy of science. Since its inception, it has dealt with the complex interrelations of natural sciences, technologies and every dimension of social structures and practices associated with them. STS can be understood as a 'reflexive empirical science', aiming to generate a better understanding of how social mechanisms and potentials shape sociotechnical development in different places, times, industries, technologies and (non-)human configurations[1].

Much work in STS puts a special emphasis on exploring how science and technology shape social orders and vice versa.

In practical applications of STS research, bottom-up processes are implemented to consider stakeholder interests in early phases of technology development. STS here adopts e.g. methods of technology assessment [4], and tools of participatory science [5] to offer not just empirical research but recommendations and even practical implementations of technology policies "from the bottom-up". This was first widely attempted during the initial boom of nano-technologies around the turn of the millennium and again (and much more successfully) a few years into the rapidly growing field of biotechnologies. Today, robotic healthcare approaches, data-driven technologies, changing manufacturing practices (e.g. additive manufacturing), and many other high-interest technologies are closely accompanied by STS research to explore and

---

[1]For further introduction into the core premises of STS, see Felt et al. ed. 2017 [3].

promote benefits that answer to societal needs and adhere to appropriate processes of decision-making.

Another focus is on science communication and public engagement in controversies surrounding science and technology. While today, a lack of trust in scientific facts and technological innovation is often perceived as a new threat by scientists and engineers, strategies of future successful incorporation of stakeholder needs with the systemic goals of science and industry have been experimented with by STS practitioners for quite some time now [6].

One key argument made by STS scholars is that communicating science and technology to society in a unidirectional way – from scientists and engineers to the public – no longer works for building trust in technologies that are inherently linked to social concerns like sustainability. Rather, dialogue and interaction between experts and public is essential for successful future innovation, as it can allow for a better tailoring of research goals and technological solutions to societal needs. Strengthening such interactions includes enabling the participation of citizens and non-experts in the conduct of research, development, operations and monitoring of relevant sociotechnical systems. Where and to what extent public are to be involved within such a setting then drastically depends on the desired outcome of their participation in the process: While, for example, increasing the quality of a certain technological solution through including lay knowledge is one possible goal, another one could be to train not only the technological but also the intrapersonal, political or civic capacities [7, p. 638] of stakeholders through interaction with scientists and engineers.

A third body of work in STS explores the networked global character of many modern sociotechnical challenges, including the question of global sustainability and ways to cope with the yet unresolved issue of how to go into a desirable future inherently limited by our planet's limited resilience to non-sustainable sociotechnical activity. For shedding light on such issues, societal discourses on a large scale can be traced and categorized to understand how a certain sociotechnical challenge – like space debris – is already and might further develop to be of importance as a sustainability challenge on a societal level.

3.3 SPACE DEBRIS AND SOCIOTECHNICAL SUSTAINABILITY CHALLENGES

Understanding space debris as a sustainability challenge is relevant for how societies and public perceive of the necessity of upstream and downstream space applications, spaceflight activities in general, the space sector as a scientific, technical and political community, and its importance and legitimacy as a project of modern societies. In this paper, we argue that space debris is increasingly understood as a sustainability challenge with orbital environments needing to be sustained and protected from further congestion.

3.3.1 Space, the finite frontier. Sustainability within limited environments

While sustainability concerns have gained considerable attention in public discourses (e.g. when it comes to sustainable energy, maritime microplastics, food production, consumer supply chains, urban and long distance mobility), the space sector has, up until recently, received less attention regarding sustainability for a number of reasons: Before the advent of New Space, spaceflight had met little public enthusiasm as its conduct had shifted from prestigious projects of the Apollo or even still the "Space Shuttle" era to emphasis on scientific exploration, military utilization and commercial applications. With the exception of some scientific missions and for several reasons, none of these activities, especially after the end of the Cold War, had sustained significant public attention towards spaceflight or the space sector as a whole – nor were they always intended to do so.

On the one hand, publicly financed missions have been built on more or less consistent funding structures already in place within the established institutional networks of the space sector. As aerospace technologies were considered key technologies within domestic economies and were also seen as granting strategic political autonomy through access to space with continued state funding, there was not always a need to invest in expensive outreach and public awareness efforts. As space research and development could, and still can to some extent, be perceived along the lines of "big science"[2], state driven "departmental research", or even a "triple helix" mode of institutional innovation [9], it very much functioned as a secluded, largely publicly funded ecosystem not only in the economic sense, but equally in its limited public perception.

On the other hand, private initiative in the space and ground segment, though proving partially successful from the 1980s onwards, did not require spaceflight activities to be a topic of much public interest either. Telecommunication and Earth observation businesses were content being part of a relatively small and clear cut (institutional) market for space applications that required neither widespread technology hype nor public interest in the activities of the space sector as a whole. As reliability of components, subsystems and systems is often identified as the paramount goal of spacecraft development, development demands like sustainable design seem to have played a secondary role so far. Redundancy and rigorous testing within systems engineering processes, e.g. the well-known and often adapted Technology Readiness Level (TRL) approach, have been perceived as the core expertise and concern of the space sector and its development processes.

Similarly, the launch sites of satellites and other payloads, which are, by nature, mostly located in remote polar or equatorial areas far outside the scope of public attention [10], have so far not sufficiently responded to calls for a more sustainable conduct of launch activity. While the use of toxic propellant by various launchers has indeed reached public debates from time to time, the space sector has not been

---

[2]For a classical analysis of big science, see De Solla Price 1963 [8].

thoroughly questioned with regards to its ecological footprint in certain regions of the world.

Lastly, while sustainability is today broadly discussed and present in respect to many planetary ecosystems and environments, the ecological quality of outer space environments has been largely neglected so far: As concerns of ecological sustainability often arise when limited and vulnerable environments are perceived as being threatened by sociotechnical (mis-)use, approaching outer space as an *environment* is a key aspect for public attitude towards the New Space era.

Space debris might become the fulcrum, at which these questions could be decided in the near future as it raises concerns for both the limited resource of available orbital slots [11] as well as the potentially inherent worth of outer space as a natural environment that should not be polluted and consumed without closely enforced regulation. Through space debris as a cumulating sociotechnical "waste" problem in Earth's orbits, outer space can be increasingly understood as a beyond-planetary environment [12] not only concerning certain strategic, scientific or commercial interests, but citizens and other societal stakeholder alike.

Increasing consensus is established, that it needs to be contained. Similar to other forms of human-made waste like nuclear waste or micro-plastics challenging local and global ecosystems as well as societal and individual health, we currently witness a shift in the ways a once accepted by-product of scientific-technological progress, economic interests and geopolitics is no longer considered sustainable.

As the "rising dependency of societies on satellites [...] increase[s] competition over space assets" [13, p. 5], this also includes available orbital space and thus space debris which limits it. Without effective management of space debris, increasingly essential satellite infrastructures are at risk.

This has recently been emphasized by ESA's director general, Johann-Dietrich Wörner, who observed that [14]:

> "Space is different from 50 years ago. Then, it was a race between superpowers; today, it is everything. We all rely on space each and every day".

Space and especially Earth's orbits, as his statement indicates, have gained political and public attention (or inattention) similar to that of fundamental planetary issues of today's concern – climate change probably featuring most prominently among them.

3.3.2   Space sustainability as a security issue

In the wake of space infrastructures having reached a critical status for the functioning of global societies, the sustainability of space assets increasingly coincides with concerns for their security. As the European Commission has recently stated: "Space technologies, infrastructure, services and data provide the EU with the tools needed to address societal challenges and big global concerns [...] The security and well-being of our citizens increasingly depends on information and services provided from space." [15, p. 5].

This novel risk configuration is the legacy of "Old Space" and its neglect of sustainability issues. Since the launch of Sputnik, outer space has become increasingly crowded with human material culture [16], functioning and non-functioning satellites, upper rocket stages, probes, landers, modules, organic human remains. At the same time, government spending on space programs and budget went down, accompanied by a lack of ability to tackle infrastructural maintenance and to sustain future visions for mankind in outer space.

Core elements of US space doctrine had been accomplished – the moon was conquered, the US American flag placed on another planet, the race against the Soviet Union won, while global hegemony extended to spaceflight capabilities. In short, the driving forces of the 1950s to 1970s Space Race and the accelerated militarization of space during the 1980s had vanished [17, p. 60]. Accordingly, formerly prestigious and ambitious undertakings soon fell short of gaining public interest and political support around the globe. Caring for the already existing satellite infrastructure or its ensuing breakdown in the form of space debris was not considered worthwhile during the post-Cold War global détente, as it would have been a reminder of the era's materialized heritage still looming above in Earth's orbits, the large sums spent on militarized competition and the multiplicity of mutual espionage infrastructures.

Even more so, since the political status of many space technologies, in particular remote sensing satellites, was and is still considered one of confidentiality – (supposedly) cleaning up another state's former assets in space has always been a sensitive and contested issue in international relations and particularly international law. This is mainly due to their often proclaimed "dual use" character; taking down another countries' defunct satellite could and still can cause severe diplomatic tensions. Knowledge of the global status of satellite infrastructures thus remained limited to a small collective of technical experts [2]. This has, for a long time, left space debris abandoned for the sake of diplomatic relations – and thus up for future generations to take care of.

If we think of "[s]ocieties [as] being 'grounded' in infrastructure; their functioning, continuity and survival are made possible by the protection of infrastructure" [18, p. 500f], then space debris might in the near future be 'grounding us' – confining humanity to Earth, and making space assets unusable due to exceeding space debris pollution.

On the level of European space policy, debris is now increasingly framed as a serious threat to humanity by policy makers and media and in need of being contained. While more than 90% of orbital debris has been attributed to the two major space race powers, the United States and Russia, as well as China due to its large-scale anti-satellite tests, the demise of ESA's flagship satellite ENVISAT in 2012 has put Europe in a new position of responsibility [19]. According to a discussion in the European Parliament, interconnected questions concerning "safety, security, and sustainability" of orbital assets make necessary "a high degree of care, due diligence and appropriate transparency" [13, p. 7]. For instance, a decision by the European Parliament and the European Council claims that "[s]pace debris has become a serious threat to the security, safety and sustainability of space activities", and that the support framework of ESA's Space Surveillance and Tracking (SST) program "should contribute to ensuring the long-term availability of European and national space infrastructure facilities and services which are essential for the safety and security of the economies, societies and citizens in Europe." [20, p. 2].

Here, the potential social and economic risk posed by space debris to European citizens under the prospect of its cascading self-destruction is evoked by framing functional space infrastructures as essential parts of functioning societies. More precisely, the perceived need for continuous and uninterrupted availability of space assets derives from their status as parts of a sensory, navigational and communicational public provision. This again, turns maintaining sustainable conditions in orbits inhabited by these infrastructures into a broader public security issue. As mega-constellations could be expected to multiply the number of active satellites in LEO if unchecked, "access to space and operations in space will continue to increase in complexity and to face growing threats, with the ultimate risk of losing capacity to explore and use space." [19, p. 1]. Creating a secure space environment for all nations requires not only technical or military solutions but deterrence and diplomacy, as well as public and political awareness of space debris as first and foremost a threat to sustainability, both on Earth and in orbit.

3.4   SPACE DEBRIS AS A BIDIRECTIONAL RISK PHENOMENON

The realization of our reliance on space infrastructures as a modern society, leads us to the second dimension of understanding the societal impact and perception of space debris. As mentioned before, space debris has the peculiar habit of facing public perception as a double risk: through harm caused by re-entry and surface impact as well as through its negative consequences for sustainable and secure satellite-based services that we regularly use. In this sense, it is not just a double risk but a bidirectional risk: As a space phenomenon, it causes unwanted effects on the planet below from above. As a technology of planetary origin, it endangers the orbits as a spatial resource and an environment only able to safely incorporate a limited number of artificial objects while remaining able to host critical satellite infrastructure.

This fact distinguishes space debris from many other sustainability issues that currently receive much attention in public discourses and policy making alike. Current and future national sustainability agendas, European framework programs like Horizon Europe [21], just as the United Nations Sustainable Development Goals [22], put emphasis on sustainability on a global scale, yet, up until now, this notion of sustainability is widely synonymous with planetary sustainability. Space debris, through its bidirectional nature, does not fit the bill: While on the one hand it *is* a planetary challenge and subjectable to perspectives of global governance for public care – no nation wants its citizens to be threatened by debris re-entry just as by any other hazardous environmental waste. On the other hand, it is *much more* than that, namely a distinctive 'outer space issue', that, while it is caused by the remains of technologies produced and launched on Earth, affects outer space as an environment; an environment that public can perceive as such, because space debris as sociotechnical by-product of spaceflight activity changes the way we look at and think about the importance of finite orbital spaces above.

The parallel occurrence of both "risk vectors" can thus be expected to play a considerable role in societies perceiving of at least LEO as an environment that is both the target and source of sociotechnical risk. While space sector experts already refer to such a "space environment" [23] on a regular basis, public perception can be expected to adopt and interpret the term based on concern for their well-being as well as for what could be perceived a limited resource worthy to be sustained out of long term societal interest.

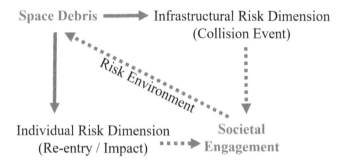

Figure 3.1: Space debris seems to define outer space as a bidirectional risk environment in public perception: as an individual risk on the surface as well as an infrastructural risk in orbit

As such, space debris rises as a sustainability challenge simultaneous to the ways in which policy actors and the public perceive of outer space as an environment worthy of protection. For instance, as the first mega-constellations were being launched, they were immediately spotted even by inattentive observers looking at the night sky [24], contributing to novel ways of how LEO becomes subject to public interest and concern. In other words, and as shown in Figure 3.1, the dual risk of space debris leads us to view outer space as a risk environment and subsequently, as a new kind of sustainability challenge. This process signifies fundamental changes in the ways societies use and relate to outer space: Not only does the increasing commercialization of outer space mark a challenging technological and regulatory endeavour, it also introduces novel structures, practices and organizational forms of exploration, exploitation and excitement.

New actors – from garage-based start-ups to large tech-businesses from a variety of countries – are currently approaching uncharted territories in space technology innovation, contributing to an enthusiastic debate about the promises of commercial spaceflight. With these new players, private activity, and new kinds of relatively low-key downstream applications entering the space business, societal awareness and relevance of the space sector as well as LEO as a valuable environment can be expected to further increase. It is likely, that increasing attention is given to space debris not only as a potential impediment to the growth potentials of the New Space business, but also as a hurdle to be overcome in allowing the future democratization of more openly accessible orbits.

3.5 CONCLUSION

Successfully tackling space debris as a bidirectional sustainability risk is thus to acknowledge two key aspects: (1) It is difficult to accomplish, requires a thoroughly collaborative international approach, and has to be recognized as a sociotechnical challenge of global magnitude that societies might not (yet) be fully equipped to engage with; (2) It is possible to identify this risk as a chance and incentive to consider the role of outer space as an environment in the public perception of and interaction with outer space and the space sector.

As with the governance of other global sustainability issues like climate change, various hurdles stand in the way of finding sufficient and satisfying answers to make our future sustainable. Often referred to challenges like that of the tragedy of the commons do also apply to sustainability concerns in outer space. Finding effective ways of international collaboration, e.g. through establishing robust institutional frameworks to enforce and incentivize a more cautious and coherent handling of the matter of space debris, are required.

However, taking all the points mentioned in this chapter together, space debris holds one decisive advantage to be effectively contained as compared to other challenges of environmental sustainability: its remote nature and the unresolved societal status of outer space as an environment. Dissimilar to the oceans, the atmosphere, the rain forest, our urban landscapes, and many other disputed and endangered environmental goods, there is still much room for us as a society to debate about how we value outer space as our planetary backyard and want to be involved in sustaining it.

This might give considerable leverage to adopt extended stakeholder participation in the process of space debris mitigation and removal, but also the future of New Space as a whole. With renewed interest in spaceflight and satellite assets currently on the rise, public participation in finding solutions to the space debris problem can and should be fostered. As STS and other related fields have frequently shown, not only does societal acceptance of technological solutions rise though meaningful stakeholder participation, but so often do their *quality*, too.

On the side of space debris mitigation, this could for example mean intensified involvement of new groups of actors in evaluating downstream demand for New Space applications, prioritization of services to be delivered via orbits, or even the mission design of satellite constellations themselves. This could provide considerable potential for improving the societal acceptance of useful New Space technologies while preventing "trial and error" approaches to technologies and business models unnecessarily leaving behind space debris.

Similarly, regular public participation provides ample opportunity to address questions such as: What levels of risk of in-orbit collisions are acceptable in relation to the value of the outer space environment? What are reasonable and justifiable thresholds for injury and damage by space debris re-entry? Which missions and applications are most valuable for societies to be provided from orbit? While these questions require strong expertise and professional experience to answer and put into policy, stakeholder involvement could bring not only an "outside" view that potentially merits previously unconsidered aspects, but it also might strengthen the legitimacy of standards and policies of truly global reach.

When it comes to Space Situational Awareness (SSA) and the handling of grounded debris fragments – to name but two examples –, similar opportunities of participation would be advisable.

One particular format of participatory science and development that has been tested over the past years is that of "citizen science". While citizen science has already to some extent been implemented in discovering Near Earth Objects (NEOs) and the detection and analysis of different astronomical phenomena, it has not yet gained widespread use in the context of space debris activities. Engaging citizen scientists not only in the tracking but also the recovery of space debris fragments – which means before and after re-entry – would be a promising approach to include public in the enterprise of managing the risks of space debris.

What can be expected as a result of public participation in engaging space debris as a bidirectional sustainability issue, could, broadly defined, be understood as socio*technical* resilience in the handling of the challenge. Such an approach would not be primarily about raising tolerance for potential harm caused by space debris in the future while seeking technological solutions without public involvement. Instead, space debris offers the opportunity to strengthen resilience through stakeholder participation in space policy making and technological development alike. As space activities – including their sustainability challenges – would then be more easily understood as "by and for" society, space debris could be tackled based on a broader and more stable societal consensus.

ACRONYM

LEO Low Earth Orbit

NEO Near Earth Object

SSA Space Situational Awareness

SST Space Surveillance and Tracking

STS Science and Technology Studies

TRL Technology Readiness Level

GLOSSARY

Kessler syndrome: The proclaimed effect of space debris collisions cascading out of control, forming a debris population that makes impossible extended future use of (certain) orbits.

Mega-constellations: Often refer to (small) satellite networks of hundreds of satellites or more, being launched at low cost with the aim of providing high redundancy and global coverage for a satellite-based service.

New Space: Is often the term used for a new and innovative approach of organizing and conducting spaceflight activities with mostly commercial interest by private actors.

Risk: Risk, in the sense discussed here, does not only refer to the (numerical) probability of an undesired effect of a certain sociotechnical system occurring, but also includes the dimension of societal anticipation of such probabilities.

Space segment / Ground segment: Space sector activities and assets are often discerned as being operated and placed in outer space (satellites), or operated and located on Earth (e.g. ground stations).

Tragedy of the commons: A concept claiming that environments and other entities of shared ownership are often faced with a lack of care and responsibility by those who share them.

Upstream applications / Downstream applications: In the space sector, upstream activities often refer to hardware and investment-heavy products, while downstream business is often associated with data-driven applications and spin-off commercialization of space assets.

FURTHER READING

Damjanov, K. (2017). *Of Defunct Satellites and Other Space Debris.* Science, Technology, & Human Values 42 (1): 166–85.

Gabrys, J. (2011). *Digital Rubbish.* Ann Arbor, MI: The University of Michigan Press.

Gorman, A. C. (2014). *The Anthropocene in the Solar System.* JCA 1 (1): 87–91.

Losch, A. (2019). *The Need of an Ethics of Planetary Sustainability.* International Journal of Astrobiology 18 (3): 259–66.

Parks, L. (2013). *Orbital Ruins.* NECSUS. European Journal of Media Studies 2 (2): 419–29.

Bibliography

[1] D. J. Kessler and B. G. Cour-Palais. Collision Frequency of Artificial Satellites: The Creation of a Debris Belt. *Journal of Geophysical Research: Space Physics*, 83(A6):2637–2646, 1978.

[2] C. J. Newman and M. Williamson. Space Sustainability: Reframing the Debate. *Space Policy*, 46:30–37, 2018.

[3] U. Felt, R. Fouché, C. A. Miller, and L. Smith-Doerr. *The Handbook of Science and Technology Studies.* MIT Press, 2017.

[4] K. Konrad, A. Rip, and V. C. S. Greiving-Stimberg. Constructive Technology Assessment–STS for and with Technology Actors. *EASST Review*, 36(3):13–19, 2017.

[5] J. Chopyak and P. Levesque. Public Participation in Science and Technology Decision Making: Trends for the Future. *Technology in Society*, 24(1-2):155–166, 2002.

[6] A. Delgado, K. Lein Kjølberg, and F. Wickson. Public Engagement Coming of Age: From Theory to Practice in STS Encounters with Nanotechnology. *Public Understanding of Science*, 20(6):826–845, 2011.

[7] C. Selin, K. C. Rawlings, K. de Ridder-Vignone, J. Sadowski, C. Altamirano Allende, G. Gano, S. R. Davies, and D. H. Guston. Experiments in Engagement: Designing Public Engagement with Science and Technology for Capacity Building. *Public Understanding of Science*, 26(6):634–649, 2017.

[8] D. J. de Solla Price. *Little Science, Big Science*, volume 5. Columbia University Press New York, 1963.

[9] H. Trischler. The "Triple Helix" of Space: German Space Activities in a European Perspective. Technical report, 2002.

[10] P. Redfield. The Half-Life of Empire in Outer Space. *Social Studies of Science*, 32(5-6):791–825, 2002.

[11] S. Durrieu and R. F. Nelson. Earth Observation from Space–The Issue of Environmental Sustainability. *Space Policy*, 29(4):238–250, 2013.

[12] L. R. Rand. *Orbital Decay: Space Junk and the Environmental History of Earth's Planetary Borderlands*. PhD diss., University of Pennsylvania, 2016.

[13] European Parliament. 2016. REPORT on Space Capabilities for European Security and Defence (2015/2276(INI)), Official Journal of the European Union (27.5.2015) (accessed on July 9, 2018). https://eur-lex.europa.eu/legal-content/EN/TXT/PDF/?uri=CELEX:32014D0541&from=EN.

[14] S. Clark. *It's Going to Happen: Is the World Ready for War in Space?* The Guardian, April 2018. Accessible on: https://www.theguardian.com/science/2018/apr/15/its-going-to-happen-is-world-ready-for-war-in-space.

[15] European Commission. 2018. Horizon 2020. Work Programme 2018-2020. 5.iii. Leadership in Enabling and Industrial Technologies – Space, Decision C(2018)4708 (accessed on May 20, 2020). http://ec.europa.eu/research/participants/data/ref/h2020/wp/2018-2020/main/h2020-wp1820-leit-space_en.pdf.

[16] A. Gorman. *Heritage of Earth Orbit: Orbital Debris–Its Mitigation and Cultural Heritage*. Ann Garrison, 2009.

[17] H. E. McCurdy. *Space and the American Imagination*. 2nd ed. Baltimore, Md., London: Johns Hopkins University Press, 2011.

[18] C. Aradau. Security that Matters: Critical Infrastructure and Objects of Protection. *Security Dialogue*, 41(5):491–514, 2010.

[19] European Space Policy Institute. 2018. Reigniting Europe's Leadership in Debris Mitigation Efforts. "ESPI BRIEFS" No. 19 (accessed on May 28, 2020). `https://espi.or.at/files/news/documents/ESPI_Brief_19.pdf`.

[20] European Parliament/Council. 2014. Decision of Establishing a Framework for Space Surveillance and Tracking Support, (541/2014/EU) (accessed on May 22, 2020). `https://eur-lex.europa.eu/legal-content/EN/TXT/PDF/?uri=CELEX:32014D0541&qid=1590134547779&from=en`.

[21] European Commission. 2019. Orientations Towards the First Strategic Plan for Horizon Europe (accessed on May 22, 2020). `https://ec.europa.eu/info/sites/info/files/research_and_innovation/strategy_on_research_and_innovation/documents/ec_rtd_orientations-he-strategic-plan_122019.pdf`.

[22] United Nations. 2020. Transforming our World: The 2030 Agenda for Sustainable Development (A/RES/70/1)) (accessed on May 22, 2020). `https://sustainabledevelopment.un.org/content/documents/21252030%20Agenda%20for%20Sustainable%20Development%20web.pdf`.

[23] European Space Agency. 2019. ESA's Annual Space Environment Report 2018. GEN-DB-LOG-00271-OPS-SD (accessed on May 26, 2020). `https://www.sdo.esoc.esa.int/environment_report/Space_Environment_Report_latest.pdf`.

[24] I. Sample. *Companies' Plans for Satellite Constellations Put Night Sky at Risk*. The Guardian, 9 January 2020. Accessible on: `https://www.theguardian.com/science/2020/jan/09/companies-plans-for-satellite-constellations-put-night-sky-at-risk`.

IV

Technological Challenges & Current Developments

Overview of the Proposals for Space Debris De/Re-Orbiting from the Most Populated Orbits

Andrey A. Baranov

Keldysh Institute of Applied Mathematics of RAS, Moscow, Russia
Peoples' Friendship University of Russia, Moscow, Russia

Dmitriy A. Grishko

Bauman Moscow State Technical University, Moscow, Russia

CONTENTS

THIS chapter surveys the available approaches to the problem of large space debris mitigation. The first part contains a brief description of the main engineering solutions that can be applied for capturing and removing large space debris objects. We consider tether systems, electrodynamic tethers, manipulators, contactless
ion-beam systems, laser systems and solar sails. The second part outlines the basic
approaches to finding flyby sequences for removing large objects of space debris from
a group to disposal orbits. We consider both low Earth orbit and the vicinity of the
geostationary orbit. De/re-orbiting can be effected in two ways which differ in the
role of an active spacecraft: it either transfers between objects (which are pushed to
disposal orbits by special modules that are accommodated on their surfaces) or de-
livers independently an object to a disposal orbit and then returns back to deal with
the next object. The third part provides a survey of both planned and implemented
projects aimed at demonstrating the possibility of objects removing to disposal orbits
or spacecraft repairing. Such an involved task should be implemented by using a collector which will perform fundamentally new functions: capturing and de/re-orbiting
objects from orbits or carrying out various service operations.

4.1 INTRODUCTION

Man-made contamination of near-Earth space is currently an urgent problem of astronautics. A collision of a spacecraft (SC) with even a small piece of space debris can
damage vital on-board systems and make the SC inoperable. The greatest danger is
posed by large space debris objects (SDO), as their collision at high relative velocities
with each other or with a healthy SC can lead to the appearance of a huge number
of small fragments, which can eventually provoke the Kessler collisional cascading
effect [1,2].

In spite of SDO mitigation measures taken on a state-by-state basis (for example, ESA's "Clean space" program [3] or Inter-Agency Space Debris Coordination Committee (IADC) recommendations [4, 5]), it is also necessary to develop systems for de/re-orbiting spent SC, launch vehicle stages and upper stages. According to the simulation results of [6–9], annual removing of ~5 large SDOs is required to prevent a cascade increase in the number of hazardous objects in low-Earth orbits.

As of today, several methods of space debris mitigation have been proposed. Active methods call for a direct transfer of space debris objects into dense atmosphere (mainly for low orbits) or their transfers to disposal orbits (DO) with the help of active SC. Passive methods involve no direct contact with space debris.

De/re-orbiting of large SDOs to DOs is a difficult engineering problem with several available approaches, which differ from each other in methods of dealing with objects (nets, harpoons, robotic arms, contactless methods) and in methods of how objects are towed to DOs (liquid propellant engines, electric propulsion engines, solar sails, unfurlable aerodynamic surfaces).

Large SDOs are mainly distributed in the following three altitude ranges: 600-1,500 km (LEO, operational range of long-term low-orbital SC), 18,000-24,000 km (MEO, operational range of global positioning systems), and the region near the geostationary orbit (GEO, 34,000-37,000 km). For each of these types of orbits, conceptual design proposals (of different developmental maturity) for SDO de/re-orbiting space missions are already available. As a rule, such approaches involve only one method of controlling an object and only one method for its removing to a DO, because an attempt to apply several concepts at the same time substantially complicates the design and control systems of the SC-collector.

4.2 ENGINEERING SOLUTIONS FOR SPACE DEBRIS OBJECTS DE/RE-ORBITING MISSIONS

A survey of possible means of large SDO capturing is given in [10]. For a contact-free object capturing, an SC is located at some distance from the object, mechanical linking is effected by special means, for example, by using harpoons or nets. For a contact capture, an SC should approach an SDO, and mechanical linking is effected between a docking unit of an active SC and some structural element of the SDO. For example, for a spent launch vehicle stage, this can be either the equipment bay structure on which the payload is accommodated or the main launch vehicle engine chamber.

4.2.1 Tether systems

One of the most promising methods of SDO removing to DO is to attach a tether to an SDO object and then tow it. Tether systems are almost independent of the shape of an SDO and the rate of its own rotation, but the tether dynamics substantially complicates the controllability of an "active SC–SDO" system. Thus, the beginning of the towing stage is associated with the process of repeated stretching and weakening of the elastic tether. As a result, the tether can transmit the towing thrust irregularly, which can lead to "swinging" of the towed object up to large amplitudes and even to a tumbling process.

Motion simulation of an "active SC-SDO" system was performed at Samara University [11]. A system consisting of a low-thrust tug, a passive object (simulated by a long solid body heavier than the tug vehicle) and an elastic tether was considered. In the model, possible rotations of the towed object and tether sagging were taken into account. By numerical experiments, the towing parameters ranges were identified for safe capturing and subsequent removing of a possible object.

Extended full-scale experiments conducted by ESA and computer simulations justified the feasibility of tether systems that involve harpoons or nets [3].

The use of harpoons allows one to capture a target at non-zero relative velocity and does not require the presence of a docking module on a de/re-orbited object. The harpoon is fired from the SC and pierces the SDO, thereby providing a mechanical link with the object. The disadvantage of this method is that at the moment of piercing the body receives a shock pulse relative to its center of mass, which can result in an additional angular velocity of the object. The harpoon approach has another disadvantage: harpoon penetration into the fuel tank of a liquid propelled stage can result in an explosion of fuel residues in the tank or the appearance of additional rotation caused by the discharge of gasified fuel residues through the hole in the stage body.

Studies on harpoon capturing of space debris objects, as conducted by EADS Astrium on bench models, show that an Astrium-designed harpoon [12] is capable of capturing targets with angular speed up to 6 degrees per second. The hitting accuracy is 8 cm at 10 m firing distance. According to experimental estimates, the harpoon system is capable of transporting SDOs of mass up to 9,000 kg, secondary debris formation being minimal and occurring mainly inside the target body.

The design of an SDO removing system should also take into account the fact that as "objects" one should consider not only entire satellites or upper stages of launch vehicles, but also their fragments. This is why SDO capture methods involving throwing nets have been proposed and studied. In this case, an SDO removal from the orbit is effected by means of a net that is fired from an active SC and is deployed using guiding loads. A net can be applied to objects of different shapes and sizes, but does not eliminate the possibility of a collision between an active SC and an object. The capture system should be accurate enough to effectively perform the process of removing with due regard of the control of the involved system for damping and vibration of flexible elements and joints.

Simulation of such a system was carried out at the Department of Aerospace Science and Technology of Politecnico di Milano [13]. An engineering solution has been developed to calculate the capture and transporting dynamics of space debris taking account the principal problems in control and navigation. For simulation it was assumed that an active SC with a net cannot be operated remotely due to the "dead zones" in orbit, and hence a stand-alone mode should be envisioned. For the experiment, a software product was developed for accurate simulation of the dynamics of deployment, contact, and closure of the net, and of the towing dynamics and transporting dynamics. The net motion was described using a discretized viscoelastic tether model.

Research aimed at development and testing of various net systems has also been conducted at the McGill University (Canada) [14]. As a result, the concept of a locking mechanism was developed based on tether expansion. The concept was implemented and tested on a laboratory bench consisting of a reduced net, a supporting frame and a space debris mock-up (derived from the second stage of Zenit-2 launch vehicle, Fig. 4.1). Test bench experiments demonstrated stable operation of the locking mechanism under Earth gravity conditions and manual locking initiation.

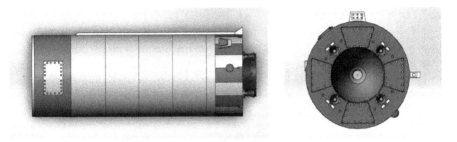

Figure 4.1: 3-D model of "Zenit-2" LV's second stage

Computer simulation of the net closure system under zero-g conditions and for a realistic SDO model confirmed the viability of the closure concept, in spite of a number of constraints adopted in the simulation.

4.2.2   Electrodynamic tether

An electrodynamic tether can be used for de-orbiting LEO objects. A difference in the potential between the tether and an SDO can be achieved by using solar panels installed on an active SC. The electrodynamic forces arising in the tether as it moves and interacts with the Earth magnetic field will theoretically change the trajectory of the SC–SDO system in order to push it into the dense layers of the atmosphere. Such an approach does not require any fuel for de-orbiting an SDO, which makes it more efficient in comparison with other engineering solutions. However, under this approach the risk of collision with other space objects is higher, because the length of such a tether is approximately one kilometre.

The EDDE (ElectroDynamic Debris Eliminator) SC is one of the prototypes of this system [15]. This SC is aimed at de-orbiting fragments of space debris of mass >2 kg from orbits of altitude 500-2000 km. The design mass of the EDDE SC is 100 kg, the required power is ~7 KW. A docking with an SDO is effected using a net. The system for SDO de-orbiting consists of the EDDE SC equipped with solar array panels (capable of accumulating large charges) and an electrically conductive tether of length ~1 km (made of a strip reinforced aluminium conductor of width 30 mm and of thickness 38 μm). The current will start to flow in the tether due to the available ion atmosphere around the SC and because of the accumulated charge in the tether. Estimates show that 10 days are required to de-orbit an object of mass 1,000 kg from a 800 km orbit, i.e., the EDDE SC is capable, within a single operation day, of de-orbiting a load comparable with its own mass.

4.2.3 Manipulators

The most accurate docking to the target is achieved by using manipulators. However, for robotic docking systems the requirements are too stringent for the accuracy of SDO detection vision systems and the analysis of the most convenient capture points. An active SC is responsible for rendezvous with an SDO and compensation of the relative velocities. In turn, the motion of the manipulator is calculated from the solution of inverse kinematics problems for each of its parts. The SC onboard computer processes data from the vision system installed on the manipulator and finds a docking point using a three-dimensional model of the object. At present, capturing of controlled spacecraft in the semi-automatic mode has already been applied at the International Space Station (Fig. 4.2, NASA).

Figure 4.2: ISS crew members grapple HTV using the robotic arm Canadarm-2 and berth it to the station (Source: NASA)

A robotic docking system is studied, for example, in [16], where an SC with two robotic arms designed for a full sequence of SDO de-orbiting is considered. An SC is launched directly into the orbit of the first object scheduled for de-orbiting. After approaching the target, the spacecraft gets in-sync with the tumbling object. Using a set of sensors and actuators, which are discussed in detail in [16], the manipulator is connected to the target object. The system features two manipulators: one is designed for capturing an identified SDO, and the other one, for installing a thruster de-orbiting kit (TDK) on it for de-orbiting purposes. Next the spacecraft is undocked from the debris and transmits an activation signal to the TDK engine. After this, the SC moves to the orbit of the next in line SDO. The above operations are repeated until all TDKs will be used.

Another robotic complex for space activities is considered in [17]. The robot consists of two robotic arms with grippers and involves various modifications depending on the type of an SDO that will be captured by the robot in a remote control mode. Under this approach, one "hand" of the robot is responsible for docking/capture and the second one is used either to install a TDK or for repair activities. It is also possible to use a space robot for orbital correction of existing satellites in order to avoid collisions.

In [18], it was assumed that SDOs should be de-orbited using a hybrid module as a propulsion unit. In particular, the engine is moved from the service platform to the target by means of a robotic manipulator. SDOs are de-orbited in several stages: rendezvous with the target, target capture, engine installation, undocking, engine ignition, and de-orbiting. Target capture and engine installation are performed using a robotic arm; if an SDO involves a nozzle, the engine is installed using a specially designed device resembling a corkscrew. Much attention is paid to the device of the rendezvous system, evaluation of angular motion and search of capture points. An adhesive mechanism will be used for capturing an object by a manipulator. For this purpose, the manipulator should involve flexible polymer electrodes and an adhesive surface. This approach allows one to work with any surfaces and does not impose stringent constraints on the accuracy of determination of the capture points. A hybrid rocket engine used in the project has high specific impulse, high safety, good thrust and control capabilities, and is relatively cheap. The adjustability is important for rendezvous manoeuvrers between the service satellite and SDO. The hybrid propulsion system involves liquid or gaseous oxidizer and solid fuel. The fuel and oxidizer have no contact, except when the engine is run. Various design solutions for hybrid engines are proposed depending on the parameters of the considered SDO.

4.2.4   Contactless laser systems

The solutions proposed above are mainly aimed at removing large SDOs. Laser exposure is considered to be one of the most efficient and feasible methods for dealing with small space debris objects. The pulsed laser method has advantages such as short response time, low cost and the ability to be reused. However, the range of a ground laser system is very limited due to light scattering in the atmosphere and the geographical location of a ground station. At the same time, a space laser system, for example that installed on the ISS, is incapable of providing sufficient energy for debris elimination.

In [19], it is proposed to combine the advantages of ground and space laser systems: a hybrid ground-space laser system is proposed. At the first stage, small space debris enters the laser radiation field of a ground station. Further, if the object is not destroyed before entering the "laser dead zone", then the second stage is required: engagement of the laser system installed on the orbital station. Mathematical and numerical simulations show that small (2 mm objects) at 800 km altitude can be removed through 1,553 laser pulses.

4.2.5   Contactless ion-beam systems

"Contactless" methods of space debris remediation (for example, ion-beam method) have also been considered by ESA [3]. The ion-beam method is based on the relative motion of an SDO due to the energy produced by an active SC. For this purpose, a high-speed ion beam is generated onboard the SC near the SDO. The ions are accelerated to high speeds, and the charge of the SC itself is neutralized by the emission of electrons by the cathode-neutralizer. A certain force develops as the beam of accelerated plasma impinges on the surface of the target; as a result, the SDO will move to a higher orbit (which is required, for example, for GEO re-orbiting). A low-thrust engine should be installed on the SC to maintain the distance between objects. With this method of SDO remediation, there is no need to develop special gripping devices, and the effect can be reached regardless of whether an object rotates or what shape it has [20]. This approach is planned for remediation (in particular, from GEO) of large objects of mass 1-2 metric tons. For example, in [21], it is assumed that onboard an ion beam shepherd satellite (operating in GEO vicinity) a high-speed ion beam is created and directed to an object in order to change its orbit without docking with it. However, the use of high-density ion beams involves certain physical and engineering issues, which are related, in the first instance, to peculiarities of force action of the ion beam on an SDO of involved geometry [22], and to deposition of SDO erosion products on functional devices of an active SC.

4.2.6   Solar sails and deployable additional aerodynamic surfaces

We should also mention solar sails (or inflatable devices of similar use) for the removal of large SDOs from near-Earth orbits. Initially, the idea of a solar sail was proposed for interorbital flights without consumption of propellant. However, for low Earth orbits, solar sails can be used as airbrakes, which increase the SC drag force in the upper atmosphere. The problem of sail unfurling has been extensively studied in the last 20 years. Among the sail unfurling experiments in space and on the Earth, we single out:

- the first successive 20 m frameless sail unfurling in space (Znamya-2 experiment), which was conducted on 4 February 1993 onboard the Progress M-15 cargo ship [23];

- the first long time flight of the IKAROS solar sail in 2010 in interplanetary space [24], where a 14×14 m frameless sail was used (0.62 m^2/kg sailing capacity).

The increase in the number of SC launches encourages the creation of compact and lightweight systems aimed at their de-orbiting from the target orbit after the end of operation. According to the so-called 25-year rule (ISO 24113:2019 [25]), an LEO SC must re-enter the Earth's atmosphere within 25 years. However, in the coming years, this requirement is likely to be tightened due to the rapid growth in the number of launched SC and due to the concerns about "mega-constellations". Particular attention should be paid to the growing number of SC of mass up to several tens of kilograms (nano and micro-satellites). Such objects are located mostly near the ISS orbit (of altitude $\sim$ 400 km, inclination 51.6°) and in sun-synchronous orbits (500-800 km altitude, 97-102° inclination) — this is explained by the ways they are launched. As a rule, micro-, nano-, and pico-satellites are not equipped with propulsion systems, and for such satellites the natural decay in the atmosphere may be far above the 25 years threshold. Therefore, a solar sail / additional deployable surface seems to be a convenient system for their de-orbiting [26]. The current competitors for solar sails are electric propulsion engines of various types – with the mass comparable to those of solar sails, electric propulsion systems provide more accurate manoeuvring [27].

4.3 SPACE DEBRIS OBJECTS (SDO) GROUP FLYBY OPTIMIZATION

4.3.1 Variants of large SDO removing to disposal orbits and transfer schemes between objects

An impetus for the study of flyby schemes between large SDOs was the publication of the Guidelines for Space Debris Mitigation by the IADC in 2007 [4]. While this document identified the problem of space debris at the interstate level, subsequent studies on simulation of the near-Earth space state showed that mere implementation of these recommendations is not enough and the removal of large SDOs from the most populated orbits is still necessary.

In spite of the many available approaches for dealing with large SDOs, one can single out two principal variants (Fig. 4.3) of their de/re-orbiting to DOs. Under the first variant, an active SC executes only transfers between objects and "docks" with them, and then the transportation to a DO is effected using special inflatable aerodynamic devices or thruster de/re-orbiting kits (TDK) accommodated on the surface of a derelict object. Under the second variant, an object is pushed to a DO by an active SC itself, which then executes a transfer to the next object in line from the DO of the previous object.

By economical considerations, one active SC-collector should be able to remove several SDOs. The exact number of objects is a priori not known and depends on the parameters of their orbits and the capabilities of the active SC. The preferable de/re-orbiting variant of the objects to DOs is also not known. Therefore, within each of the variants, one should try to get the most rational scheme of transfers between objects in a given group. The choice of a transfer scheme is crucial for enhancing the mission efficiency. By using an optimal transfer scheme one is capable, with the same resources of an SC-collector, to transfer between a larger number of objects and push them to DOs. In the first place, here one speaks about minimization of the total ΔV per a transfer with a given operation time of collector.

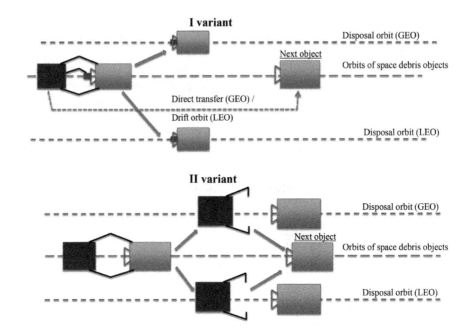

Figure 4.3: Two variants of SDO transition into disposal orbits

4.3.2 Proposals of transfer schemes for the first variant of de-orbiting to disposal orbits in LEO

In [28], transfers between 42 objects with orbital inclinations 80.5°-82° at ~850 km altitude were considered. The transfer sequence was determined by combinatorial methods (the Series Method algorithm) and the manoeuvrer parameters were found from the Lambert equation. As a result, a transfer scheme for 32 objects in 260 days with **12 km/s** total ΔV was proposed. The origin of such huge total ΔV budget is because the angles between the planes of the initial and final orbits within each transfer were corrected only at the expense of velocity impulses.

In [29], hybrid optimal control theory was applied for finding an optimal transfer sequence between LEO objects. Due to the complexity of the mathematical formalization of the problem, the solution was obtained only in the special case under the assumption that the orbits of the selected objects lie approximately in the same plane. However, in actual fact, the SDO orbits are distributed in the entire range of the Right Ascensions of the Ascending Nodes (RAAN), and so such a statement of the problem cannot be applied to real objects.

In essence, the optimal transfer problem between SDOs is close to the classical "travelling salesman problem". However, in reality it is much more complicated. Due to gravitational perturbations, the orbital planes continuously precess with different rates and SDOs travel along orbits with known orbital period. An attempt to apply traditional methods to find an SDO flyby sequence was made in [30]. It was proposed to combine the solution of the traveling salesman problem with the available orbital motion of objects taking into account the effect of the Earth polar compression. As a result, the model of [30] involved 154 binary variables, 341 real variables, and 1070 constraints for choosing a flyby sequence for 5 objects out of 11. The complexity of implementing the proposed solution increases unboundedly with the number of objects under study. In addition, the algorithm for finding solutions was not transparent and from it there was no clear reason why this or that solution was optimal.

De-orbiting of objects from Sun-synchronous orbits was considered in [16]. The mission involved the launch of an active SC, which transfers between large SDOs and carries a number of "de-orbiting devices" onboard. The authors tried to cover the following questions: the conceptual design of an active SC; characteristics of thruster de-orbiting kits responsible for pushing the object to DOs; installation of such TDKs on SDOs; transfer manoeuvres. 41 SDOs were selected with the aim removing five objects per year. The calculations showed that it is possible to de-orbit 35 objects in seven years, for which 7 launches of supply SC are required. The RAAN precession was occasionally used for calculation of manoeuvres of an active SC: for some transfers an SC was placed into a drift orbit with modified precession rate. Under this approach, no excessive total ΔV is required for changing the RAAN for lateral manoeuvres. At the same time, the choice of the drift orbit and the time spent on it were chosen in an arbitrary fashion – no relation between the desired correction of the SC orbital parameters and the parameters of the chosen drift orbit were given. The other disadvantage of this approach is that the number of de-orbited objects was assumed to be fixed per year.

De-orbiting of upper stages from Sun-synchronous orbits was also considered in [31]. A portrait of the evolution of deviations of the RAANs was used to visualize the relative motion of orbital planes (Fig. 4.4).

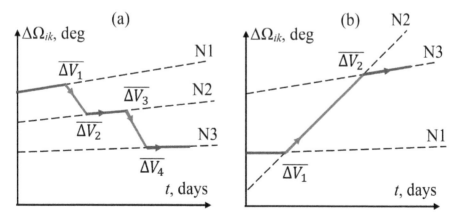

Figure 4.4: RAAN deviations' evolution portrait: successive (a) and diagonal (b) flyby schemes

A transfer of an active SC into a drift orbit with a changed natural precession was used to offset the difference ($\Delta\Omega$) in the RAAN (Fig. 4.4a and 4.5). This approach showed good results, because within an acceptable time (several months) it proved possible to make the planes nearly identical (with initial difference in the RAAN $\Delta\Omega \leq 15°$). Transfers with larger $\Delta\Omega$ were also required, but for such transfers a longer period of staying in the drift orbit was needed. Unlike previous studies, the parameters of the drift orbit were not chosen arbitrarily, but were calculated depending on the value of $\Delta\Omega$ and transfer duration between objects.

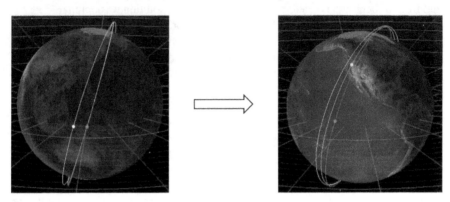

Figure 4.5: A transfer to the next SDO using a lower drift orbit with another RAAN precession rate

The manoeuvre parameters were determined on the basis of the solution of the LEO near-circular rendezvous problem, which was obtained by the authors for transfers with large RAAN deviations. In addition, for Sun-synchronous orbits the authors proposed an original transfer scheme (Fig. 4.4b) in which the orbit of the previous object is used to change the plane orientation for the transfer to the next object. This approach is called diagonal in [31]. With this complex application of the drift orbit and the "diagonal" transfer scheme, the results of [16] were **doubly** improved: it is possible to de-orbit 46 objects with the use of 4 SC-collectors.

4.3.3 Proposals of transfer schemes for the second variant of de-orbiting to disposal orbits in LEO

In [32], a constellation of 38 SC in low-Earth noncoplanar orbits was considered. It was assumed that SC can execute manoeuvres and that each of them is capable of de-orbiting 25 hazardous objects. The Lambert equation was used to estimate the manoeuvre parameters. However, in calculations no account was given of the fact that the orbits of objects undergo considerable perturbations due to the noncentrality of the Earth's gravitational field. Therefore, the need to correct the orbital planes orientation during each transfer was not considered. In addition, the possibility of removing up to 25 objects by each SC is clearly excessive and was never implemented by the authors in the solution given in the paper.

In [33] it is proposed to use an active SC equipped with a chemical or an electric propulsion system for objects de-orbiting. The transfer problem between objects in orbits with inclinations of 71° and 74° was considered. Estimations for the duration of staying of the disposal orbit's apogee in the zone of SC operation were obtained. Calculations show that the apogee of the elliptical disposal orbit for the SDO groups under consideration descends below the 700 km altitude in about 10 years. The tight time frames of the mission and stringent restrictions on the number of de-orbited objects resulted in too great total ΔV budget.

The paper [34] combines the first and second variants: the first object is de-orbited using a special module accommodated on the active SC, the next object is de-orbited by the SC itself; this approach is more reliable, but requires multiple repetitions.

4.3.4 Complex solution for determination of SDO transfer sequence in LEO

In [35, 36], the transfer problem for five SDO groups (launch vehicle last stages in LEO) was considered for both de-orbiting variants. In this case, the orbits of objects forming a group have, as a rule, the same inclination (up to several tenths of a degree).

These papers are markedly different from all other publications on SDO transfer logistics in LEO. First, they cover the main groups of large derelict objects of this type, and second, in these studies a fundamentally different approach was applied for determining the transfer sequence between the objects (this approach is simpler, but it is also precise at the same time). While most publications on this topic are aimed at building a comprehensive mathematical model and solving the problem of global optimization (for example in the framework of the "weighted traveling salesman" problem), the idea of the above approach is that the evolution of orbits itself is capable of removing the most costly corrections associated with orbital plane changes. In [35, 36], a natural perturbing factor controlling the SDO motion dynamics is singled out and used. No constraints on the number of objects under investigation are placed. The solution is obtained directly from the analysis of the portrait of evolutions of the RAAN deviations (Fig. 4.6).

Such a portrait makes it possible to simply determine the sequence of transfers for a chosen group of passive objects with near equal inclinations of orbits. This minimizes the total ΔV costs, since the most expensive corrections of orbital parameters (RAAN changes) are excluded, and in addition, the visualization of the obtained solution is provided. The authors do not fix the number of objects that should be de-orbited annually; this makes the solution more flexible. At the same time, the average rate of removal of 5–6 objects per year is achieved. As a rule, the duration of a mission does not exceed 10 years.

4.3.5 Proposals of transfer schemes for SDO re-orbiting to disposal orbits in GEO

The specifics of this type of orbits should be taken into account when constructing SDO transfer schemes in the GEO vicinity. Namely, GEO objects undergo perturbations due to the noncentrality of the Earth's gravitational field, gravitational perturbations from to the Moon and the Sun, tidal forces and Solar radiation pressure. As

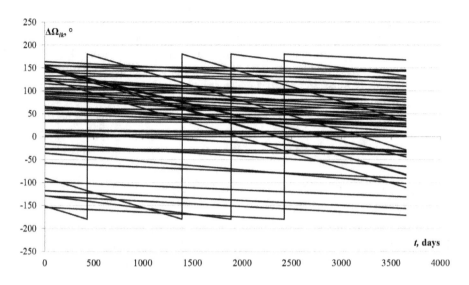

Figure 4.6: RAAN deviations' evolution portrait in the case of one space debris group in LEO

a result, once the control on the object is lost, its orbit ceases to be equatorial and geosynchronous. On the other hand, there is no upper atmosphere near GEO and the RAAN precession is about 32–35 times weaker than that for low orbits.

Alfriend [37] was one of the first to consider the transfer problem between several GEO objects within the framework of the related spacecraft service problem. The minimization of the spent fuel was taken as an optimality criterion, the transfer sequence was determined based on the solution of the traveling salesman problem. Taking into account the smallness of the coplanar components of transfer manoeuvres, the analysis was carried out only in terms of lateral components. The paper [37] was mostly concerned with small inclination orbits. Transfers of equal duration and transfers with equal fuel costs of coplanar components of manoeuvres were investigated.

The GEO refuelling problem was examined in [38]. It was assumed that a servicing SC and a fuelling station are based near a GEO and that the refuelling is performed not only by a servicing SC, but also by other usual spacecrafts with sufficient fuel budget. The numerical enumeration method and the method of particle swarm were used for determination of a transfer sequence and mass properties of an SC.

In [39], a similar problem was considered with optimization in two criteria (minimum of fuel and maximum of mission duration) and with the use of the method of particle swarm. Escape times from a waiting orbit, and the transfer durations between the waiting orbit and the target, as well as the transfer duration from one target to a different target were determined by solving the optimization problem. The Lambert equation was used for determination of the parameters of transfer orbits.

In spite of the similarity in the statements of the servicing problems, their solutions, as obtained by various authors, cannot be applied to the problem of re-orbiting of large SDOs, because these approaches involve manoeuvres back to the service station and, as a rule, zero inclination of the orbital plane to the equator. For GEO re-orbiting, it is also required either to execute direct transfers between the objects (1st re-orbiting variant) or to push the next in line object to a DO and transfer to the next object from the DO (2nd re-orbiting variant). Moreover, the transfer problem between GEO objects is highly non-coplanar.

As distinct from the case of low orbits [16, 30] and [35, 36], little attention has been paid to orbital dynamics problems for transfers between large GEO objects, and in particular, to optimal transfer schemes between such objects. One of the most recent studies is the paper [40] in which re-orbiting is carried out by the SC-collector itself (the number of active SC varies). An optimal transfer sequence is searched using a hybrid-optimal control capable of providing a sequence of double-pulse manoeuvres with simultaneous optimization of total ΔV losses and the duration of transfers. Only a small number ($n \leq 6$) of objects to be re-orbited was considered due to the complexity of the mathematical model. Similar principles were also used in [41], but in this paper it was assumed that the orbits of all objects are equatorial and circular, which however is far from reality.

In [42], both SDO re-orbiting variants from the GEO vicinity are compared (87 upper stages are considered as re-orbited objects). It is shown that due to slow evolution of the orbital parameters in the GEO region, the same transfer schemes can be used for these two re-orbiting variants. Geometrical peculiarities of relative positions of orbits in the near-equatorial region were described and two transfer schemes between objects were considered. Under the first scheme, a transfer between orbits is executed when the orbits have the same inclination near the equator, and under the second scheme, when the orbit of the next object has the smallest inclination near the equator (Fig.4.7).

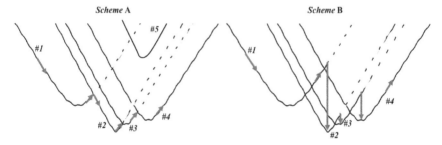

Figure 4.7: Possible SDO flyby schemes in GEO vicinity based on inclination dynamics with time

Calculations show that both flyby schemes are practically equivalent in terms of the averaged total ΔV of the transfer (between two objects) and the duration of transfers between all objects. However, not all objects under consideration can be covered under the first scheme. Hence, priority should be given to the approach when the transfers are executed when the orbital inclination of the next object attains its smallest value in equatorial region. Calculations show that 6 SC-collectors are required to clean the GEO protected region from upper stages. The active service life of each of the active SC is expected to be at most 8 years, the required total ΔV budget is at most 0.7 km/s. A conclusion is made that a re-orbiting of one object to a DO requires 10 m/s on the average; the return to a new object from the DO of the previous object is energetically nearly equal to a sequential transfer between these objects. In this regard, as distinct from low orbits (in which it is preferable to use TDKs), it is more beneficial to follow the second variant for re-orbiting space debris objects from the GEO region (i.e., using the SC-collector itself).

4.4 SURVEY OF PROJECTS BY SPACE COMPANIES AND AGENCIES IN THE FIELD OF LARGE SPACE DEBRIS REMOVAL TECHNOLOGIES

4.4.1 Orbital service problem: a key vector for ADR technology development

The design problem of transfer schemes between space debris objects is close to spacecraft service problems. Until now, orbital service of space hardware was carried out only by crews with the aim of maintaining the performance of orbital stations. One can also mention several NASA missions for inspection and repair purposes of the Hubble Space Telescope using Space Shuttles (1993–2009). The desire to reduce the costs of launching and operating a particular satellite in combination with the appearance of new technologies, has led to the idea of SC orbital servicing. Under this approach, no direct involvement of astronauts is initially planned. This is why the level of capabilities of robotic tools is a key factor governing the feasibility of a specific project.

The orbital service problem always involves a "client" (an object that requires attention) and a "mechanician" (an SC equipped with hardware, using which the client can restore its performance capabilities). Regardless of whether the client is an active satellite or an SDO, a mechanician is injected in the vicinity of the client, where it performs rendezvous manoeuvres and docks with the mechanism. At present, there are several real successfully functioning robotic space applications: Canadarm-2 (the Canadian Space Station Remote Manipulator System) and, in the coming years, the European Robotic Arm. Both of these tools are designed to be used on the ISS primarily to move cargo along its surface. When docking for a service task, a manipulator is required to capture a much smaller object and, if necessary, perform local operations on its surface, which results in smaller mass and sizes of the manipulator. Therefore, such a manipulator can be mechanically integrated into an active servicing SC.

The development of new spacecraft, as well as the management of the corresponding service robots can be carried out by simulating the operation of these systems and their physical properties in various software packages. This will make it possible to study spacecraft for outer space service purposes, to analyse the design and structures of SC and robotic systems, and to plan and implement orbital maintenance missions. The construction of a full system of operations is greatly hindered at present by the nature of the operational environment of SC. At present, the SC service strategy assumes only the monitoring and control of operation regimes of onboard systems. The abilities of ground control equipment to maintain and resume SC performance are limited by the capacities of telemetry systems, the degree of redundancy of the onboard equipment, and the available fuel in the propulsion system. It should be also noted that modern SC are nonrepairable and that their end-of-life phase is accompanied by man-made pollution of space. The root causes of failures can be subdivided into three groups: deployment failures of stowed structure elements and the impossibility of establishing a communication link with a SC directly after its injection into the orbit; onboard equipment failures; mechanical failures and damages of SC structure elements [43]. So, space debris objects are composed, in particular, of SC with failed blocks of onboard equipment but with healthy propulsion system and of SC with a healthy onboard equipment but which cannot perform nominally due to fuel depletion. The end of useful life of an SC due to fuel depletion is a typical situation for high-altitude artificial Earth satellites, and in particular, for telecommunication GEO satellites. For example, the Kepler space telescope (launched for search of exoplanets) ceased its operations because of fuel depletion.

An increase in the nomenclature of services provided by a space segment of information systems and an increase in demand of such service increase the risk of financial losses in the case of SC failures. For example, a failure in a navigation SC may result in financial losses of many dozens of large customers. In this regard, the SC service rate return becomes predominant. In such conditions, a quick repair of an SC and recovery of its health become profitable. When changing to groups of many satellites, the degree of man-made pollution of the near-Earth space increases, which requires the use of special spacecraft for removing space debris.

In other words, **no money on maintenance and repair results in money spent for space debris control**.

Service SC equipped with propulsion systems can be used both in service operations and for the removal of large SDOs to disposal orbits. We mention the following programs involving such SC (the list is not complete): ROTEX (Robot Technology Experiment), ESS (Experimental Servicing Satellite), Orbital Debris Defence System (ODDS), SUMO (Spacecraft for the Universal Modification of Orbits), Front–end Robotics Enabling Near–term Demonstration (FREND), DEOS, e.Deorbit, and RemoveDEBRIS.

4.4.2 European service mission projects

In 1993, DLR conducted the ROTEX (Robot Technology Experiment) flight experiment in the framework of the STS-55 "D-2" Spacelab mission. The objectives of the experiment were to test a 1 m robotic arm with six degrees of freedom in space flight conditions, to test communication links with the robotic arm, and to check the possibility of docking with an SC for its maintenance [44]. During the flight, the robot had to mount the truss, catch free-flying objects, connect/disconnect the power source supply in standalone mode and on commands from the ground stations. A sophisticated multisensory gripper device was equipped with 9 laser ranging system and stereo cameras. The experiment confirmed that with the current level of development of hardware and technology, robotic structures are capable of working in outer space both under the control of operators (located on the Earth or on an SC) and in the stand-alone mode [45].

The DLR ESS (Experimental Servicing Satellite) project involved laboratory experiments to study the system's dynamics, as well as to confirm the feasibility of docking of a free-flying SC with a serviced object for rendezvous, inspection, and repair by using a robotic arm [46]. The dynamics of the behaviour of an SC equipped with a manipulator and of a bundle of two SC connected by a manipulator was studied. The experiments verified the correctness of the adopted engineering solutions and enabled one to build dynamic models of the system.

The ROGER (RObotic GEostationary orbit Restorer) concept was proposed by Astrium [47–49]. A service SC of mass 3,500 kg has the ability to inspect the serviced object, stabilize and move it in other orbits using a capture system in the form of a net ejected at a distance that excludes any collision of the service SC and the serviceable SC. The reusable system was primarily designed to remove noncooperative objects from target orbits to DOs. Up to twenty ejectable nets can be used. The weight of the target capture mechanism was 9 kg, the net had four additional counterweights. Capture experiments were carried in zero-g conditions.

A similar QinetiQ's ROGER concept is based on the use of the basic design of GEO satellites. An SC of 1,450 kg launch mass (on a geotransfer orbit) has the shape of an octagonal prism and is equipped with a telescopic rod with the payload gripper system. As in the previous concept, it is proposed to use an electrojet propulsion system involving four stationary plasma engines for GEO injection and 24 hydrazine engines. The telescopic rod is mounted directly on the SC. Four "tentacles", which are placed at the ends of the rod, can operate both simultaneously and separately. The "tentacles" consist of conical fingers with soft surface on the contact side.

In 2012, Astrium also proposed the new DEOS project of an orbital service mission (of cost about 13 million Euros). The aim of the project was to demonstrate the technology of controlled de-orbiting of a derelict satellite and to perform its maintenance tasks (in particular, refuelling). To a large extent, the DEOS project depended on the technologies that had not yet been tested for space operations. The mission anticipated launches of two SC into an orbit of 550 km altitude. The "client" acted as a satellite requiring repair or disposal, and the "mechanic" had to perform necessary activities on the "client". The SC were expected to be ready for launch in 2018, but the project was closed by 2017.

In 2014, ESA began development of the e.Deorbit mission with the purpose of de-orbiting the unhealthy Envisat SC, which is the largest remote sensing satellite launched into space. The mission calls for a launch of an SC of mass 1,600 kg into an 800-1,000 km altitude orbit. The SC should approach the Envisat and capture it either by using a net or manipulators (in the latter case the object should be captured and fixed). The next step is to arrange a controlled descent of the Envisat SC into the dense atmosphere. The launch is set for 2023; the preliminary design of the mission was approved in 2016.

In 2018, an in-orbit active debris removal mission (RemoveDEBRIS) was launched. The RemoveDEBRIS, which was built by Surrey Space Centre, is based on SSTL avionics. The purpose of the mission was to demonstrate various technologies for de-orbiting space debris and to study their efficiency. The mission consists of a microsatellite platform (chaser) that ejects 2 CubeSats (targets). The SC has mass 100 kg and sizes 65 cm × 65 cm × 72 cm. It is equipped with net capture, harpoon capture, laser ranging system, and dragsail. The following experiments are foreseen:

- Net experiment. One of the CubeSats should inflate a balloon to imitate a space debris fragment. A net from the platform is then ejected to capture the balloon when it is close to the main SC. Then a de-orbiting manoeuvre should be executed.

- CubeSat-2 (DebrisSat-1, say) is ejected from the platform, after which the platform executes a series of to-and-fro manoeuvres to get data and images from LiDAR and optical cameras.

- Harpoon experiment. The harpoon scenario uses a deployable target that extends outwards from the platform and is used as a target for the harpoon.

- Dragsail demonstration. The dragsail demonstration is undertaken last when the platform is to be de-orbited. A large dragsail should be unfurled to substantially increase the aerodynamic, in order to de-orbit the SC in the dense atmosphere.

In September 2018, the possibility of using a net to capture a pre-deployed target was successfully demonstrated. In February 2019, an experiment with a harpoon was conducted: the harpoon was fired at 20 m/s speed and successfully pierced the target, which was placed away from the platform by a 1.5 m deployable boom. The LiDAR experiment was also successful, but the final stage of the mission failed. The dragsail was not unfurled and the reason is unknown.

4.4.3 USA service mission projects

The Orbital Debris Defense System (ODDS) project, which was developed at West Virginia University, is aimed at dealing with SDOs of a wide range of sizes, orbits and velocities. The spacecraft has a modular structure and is designed to meet and capture SDOs ranging in size from 10 cm to 2 m. It was assumed that it could be injected into both polar and equatorial orbits, including retrograde orbits with inclination 110°. Taking into account the fuel consumption for changing orbits and inclinations, it was envisaged to use an orbital group of such SC that operate simultaneously in the neighbourhood of several base orbits. In this case, the "neighbourhood" was understood to be a space region ranging up to 100 km in altitude and 100 km in the lateral direction from the orbital plane with respect to the base orbit.

The SUMO (Spacecraft for the Universal Modification of Orbits) project was developed by DARPA. It was supposed to launch four SUMO satellites to base near-Earth orbits. With the help of its propulsion system, an SC should approach the target-satellite to 100 m or closer, after which the system of 20 onboard video cameras will take over the control of the rendezvous with the target-satellite at the 1.5 m distance. Further docking is carried out automatically by a robot manipulator. In April 2005, tests were conducted with a six-degree-of-freedom robotic arm equipped with additional video cameras and a pulsed xenon lamp to illuminate the target and with a SC mock-up developed from the Boeing 702 GEO and Lockheed Martin A2100 space buses.

The Front-end Robotics Enabling Near-term Demonstration (FREND) project was proposed by DARPA to create a fully autonomous docking capability with satellites whose designs did not originally envisage service operations. Systems of autonomous rendezvous and capture of GEO SDOs were developed for the purpose of service and "evacuation" of SC of any type; the main working element of the system is the manipulator [50]. Full-scale laboratory demonstrators of 1 m- and 2 m-robotic manipulators were successfully tested in 2007; the reliability of the control algorithms and visualization of the robotic manipulator were checked in the fully automatic mode [51,52].

DARPA in 2007 implemented the comprehensive Orbital Express Advanced Technology project, which was to demonstrate various technologies required for autonomous in-orbit service of SCs. The Orbital Express program was supposed to change the traditional understanding of space activities by demonstrating the possibilities of in-orbital refuelling, upgrading and extending the service life of SC. The system consisted of two satellites: the ASTRO SC (for service operations) and the NEXTSat SC (a prototype of a modular serviceable next generation SC). The mission included the following operations: rendezvous in orbit, approaching, hovering near each other, capture, docking, hydrazine transfer, and replacement of special service modules. Refuelling was completed successfully. Using a manipulator, an active SC autonomously captured a free-flying "client" and transferred the battery and a module with onboard computer.

In NASA, the SC service is handled by a special unit, SSPD [53]. A description of the approaches to in-orbital SC service up to 2010 can be found in the proceedings of the conference [54]. The Restore-L automated service technology demonstrator for refuelling of the Landsat-7 SC after 2020 is currently being developed [55]. The SC is based on the Space Systems Loral SSL-1300 bus; its launch mass is from 5500 to 6,700 kg. The DARPA Agency is developing an SC for servicing GEO satellites based on technologies developed in the Orbital Express mission. The mission anticipates the solution in 2020-2021 of service tasks by means of a single SC [56].

The method of in-orbital servicing of several SC, as proposed by the Space Infrastructure Services company (USA) [57], anticipates the use of an automatic SC, which performs the above tasks of service by moving in the 2–4 years period between 5 Intelsat communication satellites. Other projects for servicing similar satellites can be found in the survey prepared by Intelsat [58,59]. The project of the MDA Corporation company [60] is primarily concerned with SC refuelling by an active SC carrying up to 2,000 kg of fuel. These prototypes of service SC are designed for providing service to satellites that are already in operation and which were not initially designed for servicing; such satellites are devoid of docking stations, are noncooperative objects, and are unfit for refuelling and replacement of units.

4.4.4 Japan service mission projects

The Orbital Debris Defense System (ODDS) project, which was developed at The Japan Aerospace Exploration Agency (JAXA) implemented the following missions and tested technologies that can be applied for in-orbit service problems:

- MFD (Manipulator Flight Demonstrator) in August 1997;

- ETS-VII (Engineering Test Satellite) in November 1997;

- JEMRMS (Japan Experiment Module Remote Manipulator System on the ISS);

- HTV (Transfer Vehicle for ISS) in 2009-2012.

The project [61] considers the possibility of using a microsatellite to de-orbit LEO space debris. In this project, an electrodynamic tether between the SC and an SDO will be used. The microsatellite docks with an object by using a manipulator equipped with a vision system that compares the detected object with the reference one and analyses the possibilities of docking. The microsatellite is navigated via GPS. It is assumed that in the future JAXA will use special labels on new SC to simplify docking of new microsatellites with an SC and their further disposal.

4.4.5 Solar sail missions: the available technology for end-of-life passive de-orbiting

At present, three successful projects have been implemented in which atmospheric braking by sails was tested for de-orbiting small SC: NanoSail-D2 [62,63], LightSail-1 [64], and LightSail-2 [65].

The main purpose of the NanoSail-D2 mission in 2010 was to achieve developmental verification of the sail deployment mechanism; the use of the same sail as an atmospheric brake for SC de-orbiting was a secondary task. The sail consisted of four petals forming a square of area ~ 10 m^2. It was assumed that due to atmospheric braking, the SC would descend from an initial circular orbit of 650 km altitude and burn in the atmosphere in 70–120 days. However, as the angular motion of the sail was uncontrollable, the SC began to rotate chaotically. The effective area of the midship was significantly less than the nominal value and the de-orbiting of the SC from the orbit took 240 days.

In 2015, the test flight of the LightSail-1 (LightSail-A) SC for the Solar Sail Demonstration mission occurred. The SC deployed its solar sail and then successfully completed the test flight in 7 days [64].

The LightSail-2 SC was launched in June 2019 [65] with the aim of demonstrating the possibility of steering the solar sail. The task was to raise the apogee and lower the perigee of the orbit by controlling the orientation of the sail relative to the Sun. It is planned that in about a year the perigee of the orbit will reach the atmosphere level, which will lead to a rapid de-orbiting.

Among the developed hardware, we also mention CubeSail nano-solar sails [66] and their analogues which are planned to be accommodated on the already launched satellites for their future de-orbiting after the end of the service life.

4.5 CONCLUSION

The practical significance of applied scientific research on the problem of space debris mitigation is confirmed by the currently existing projects for developmental verification of engineering solutions for de-orbiting of passive objects from orbits. These projects are funded by EADS Astrium, SSTL, ESA, NASA, and DARPA.

Due to funding issues, many such projects were closed, but it happened at the stages close to the beginning of practical implementation. The fact that such initiatives are seriously discussed at the level of space agency management suggests that sooner or later certain engineering solutions will be developed and tested.

To date, the principal focus has been placed on large space debris objects such as derelict satellites, launch vehicle stages, and upper stages. First, such objects are observable from the Earth (even GEO-objects), their trajectories are known and constantly refined; hence collisions with them can be avoided by accurate planning and forecasting already at the stage of selection of the operating orbits. Second, large objects pose the most serious danger in the context of the Kessler collisional cascading effect: collision of such objects results in a significant number of new SDOs with highly unpredictable trajectories.

Approximately 11% of the currently catalogued objects are fragments formed during the spacecraft's lifecycle. The amount of debris generated during space hardware operation should be minimized. At present, space agencies undertake measures to prevent the emergence of such objects. Manufacturers of modern satellites generally avoid intentional debris generation, because such debris will remain near the SC and can pose danger to it.

Current international agreements for LEO and GEO stipulate that large SDOs should be de-/re-orbited to disposal orbits at the end of their service life. However, such rules cannot be applied to objects that have become space debris prior to the adoption of these recommendations. Therefore, it is necessary to develop methods for removing such objects to safe orbits.

The topic of large-scale space debris is undergoing very rapid and vigorous development, especially taking into account the consequences of the collision of two satellites in 2009 and the anti-satellite weapons tests in near space. All large objects are catalogued and their orbits are constantly monitored. In particular, the directions related to the development of flyby methods of such objects and the methods of their capture and fixation are actively studied. The next logical step is the design of an SC-collector featuring fundamentally new functions: capture and de-/re-orbiting of objects from orbits. These solutions will certainly find their applications in the industry, and in particular, in the future field of SC servicing.

ACRONYM

DO Disposal orbit

ESA European Space Agency

GEO Geostationary Earth Orbit

IADC Inter-Agency Space Debris Coordination Committee

ISS International Space Station

LEO Low Earth Orbit

MEO Medium Earth Orbit

NASA National Aeronautics and Space Administration

SC Spacecraft

SDO Space debris object

SSO Sun-synchronous orbits

TDK Thruster de/re-orbiting kit

RAAN The Right Ascension of the Ascending Node

GLOSSARY

Low Earth Orbit (LEO): Spherical region that extends from the Earth's surface approximately up to an altitude of 2,000 km.

Medium Earth Orbits (MEO): Spherical region that extends from the average altitude of 2,000 km above the Earth's surface up to the geostationary orbit. When speaking about space debris, this term usually means the region between 19,000 and 24,000 km altitudes, where GPS, GLONASS, Galileo and BeiDou navigation systems are deployed.

Geostationary Earth Orbit (GEO): Unique equatorial circular orbit, whose altitude over the equator is 35,786 km and orbital period matches Earth's rotation on its axis, 23 hours, 56 minutes, and 4 seconds (one sidereal day). A satellite in this orbit keeps the same position in the sky relative to observers on the surface of the Earth.

Large space debris: Any passive object (non-operational satellite, last stage of launch vehicle or upper stage) that stays undestroyed after its end-of-life and can be removed to a disposal orbit.

Disposal orbit: In the case of LEO - an orbit that exists 25 years and decays in the dense atmosphere. DOs can be circular and elliptical. For last stages of launch vehicles from 850 km altitudes, for example, the approximate altitude of a circular DO may be 540-550 km, and the pericentre altitude of an elliptical DO may be 420-440 km. These values depend on area-to-mass ratio of the object, type of the orbit and the density dynamics of upper atmosphere. In the case of GEO - an orbit at least 235 km higher than the ideal geostanionary orbit.

IADC Guidelines: Document created by the Inter-Agency Space Debris Coordination Committee (IADC) and updated in 2007 - 'Space Debris Mitigation Guidelines'. Usually accompanied by 'Support to the IADC Space Debris Mitigation Guidelines' (2007). These documents describe the recommendations to space agencies, satellite production companies and launch operators and are aimed at a decrease of artificial space debris generation while carrying out space operations.

RAAN: The Right Ascension of the Ascending Node. An angle in the equatorial plane between the inertial axis Ox (now in J2000 coordinate system) and the point of orbit's ascending node.

RAAN drift: The RAAN precession (for orbits with inclination less than 90 degrees - in west direction) caused by non-spherical mass distribution inside the Earth: the shape of the planet is close to an ellipsoid of revolution and the mass distribution mostly repeats it.

FURTHER READING

Baranov A.A., Grishko D.A., Mayorova V.I. (2015). *The features of constellations' formation and replenishment at near circular orbits in non-central gravity fields.* Acta Astronautica 116:307–317.

Baranov A.A. (2016). Spacecraft maneuvers in the vicinity of a circular orbit. Moscow: Sputnik+, 512 p. [In Russian]

Labourdette P., Baranov A. (2002). Strategies for on-orbit rendezvous circling Mars. Advances in the Astronautical Sciences 109:1351-1368.

Anderson P.V., Schaub H. (2014). Local debris congestion in the geosynchronous environment with population augmentation. *Acta Astronautica* 94:619–628.

Bonnal C., Ruault J.-M., Desjean M.-C. (2013). *Active debris removal: Recent progress and current trends .* Acta Astronautica 85:51-60.

RemoveDebris Mission: Description and results (2020). *RemoveDebris Mission.* https://directory.eoportal.org/web/eoportal/satellite-missions/r/removedebris

Guglielmo S. Aglietti et al. (2020). The active space debris removal mission RemoveDebris. Part 2: In orbit operations. *Acta Astronautica* 168:310-322.

Bibliography

[1] Donald J. Kessler and Burton G. Cour-Palais. Collision frequency of artificial satellites: The creation of a debris belt. *Journal of Geophysical Research: Space Physics*, 83(A6):2637–2646, 1978.

[2] D. J. Kessler. Collisional cascading: The limits of population growth in low Earth orbit. *Advances in Space Research*, 11(12):63–66, 1991.

[3] Kjetil Wormnes, Ronan Le Letty, Leopold Summerer, Rogier Schonenborg, Olivier Dubois-Matra, Eleonora Luraschi, Alexander Cropp, Holger Krag, and Jessica Delaval. ESA technologies for space debris remediation. In *6th European Conference on Space Debris*, volume 1, pages 1–8. ESA Communications ESTEC, Noordwijk, The Netherlands, 2013.

[4] IADC space debris mitigation guidelines. `https://www.unoosa.org/documents/pdf/spacelaw/sd/IADC-2002-01-IADC-Space_Debris-Guidelines-Revision1.pdf`, 2007.

[5] United Nations office for outer space affairs. Space debris mitigation guidelines of the Committee on the peaceful uses of outer space. `https://www.unoosa.org/pdf/publications/st_space_49E.pdf`, 2010.

[6] J.C. Liou and N.L. Johnson. A sensitivity study of the effectiveness of active debris removal in LEO. *Acta Astronautica*, 64(2-3):236–243, 2009.

[7] J.C. Liou, N.L. Johnson, and N.M. Hill. Controlling the growth of future LEO debris populations with active debris removal. *Acta Astronautica*, 66(5-6):648–653, 2010.

[8] J.C. Liou. An active debris removal parametric study for LEO environment remediation. *Advances in Space Research*, 47(11):1865–1876, 2011.

[9] H.G. Lewis, A.E. White, R. Crowther, and H. Stokes. Synergy of debris mitigation and removal. *Acta Astronautica*, 81(1):62–68, 2012.

[10] V.I. Trushlyakov and E.A. Yutkin. Overview of means for docking and capture of large-scale space debris objects. *Omskii nauchnyi vestnik, Omsk Scientific Bulletin*, (2):56–61, 2013.

[11] V.S. Aslanov, A.V. Alekseev, and A.S. Ledkov. Harpoon equipped space tether system for space debris towing characterization. *Trudy MAI*, (90):21, 2016 [in Russian].

[12] J. Reed, J. Busquets, and C. White. Grappling system for capturing heavy space debris. In *2nd European Workshop on Active Debris Removal*, pages 18–19. Centre National d'Etudes Spatiales Paris, France, 2012.

[13] Riccardo Benvenuto, Samuele Salvi, and Michelle Lavagna. Dynamics analysis and GNC design of flexible systems for space debris active removal. *Acta Astronautica*, 110:247–265, 2015.

[14] I. Sharf, B. Thomsen, E.M. Botta, and Arun K. Misra. Experiments and simulation of a net closing mechanism for tether-net capture of space debris. *Acta Astronautica*, 139:332–343, 2017.

[15] Jerome Pearson, Eugene Levin, John Oldson, and Joseph Carroll. *Electrodynamic debris eliminator (EDDE): design, operation, and ground support*. 2010.

[16] Marco M. Castronuovo. Active space debris removal – a preliminary mission analysis and design. *Acta Astronautica*, 69(9-10):848–859, 2011.

[17] Hiroshi Ueno, Steven Dubowsky, Christopher Lee, Chi Zhu, Yoshiaki Ohkami, Shuichi Matsumoto, and Mitsushige Oda. Space robotic mission concepts for capturing stray objects. *The Journal of Space Technology and Science*, 18(2):2_1–2_8, 2002.

[18] L.T. DeLuca, F. Bernelli, F. Maggi, P. Tadini, C. Pardini, L. Anselmo, M. Grassi, D. Pavarin, A. Francesconi, and F. Branz. Active space debris removal by a hybrid propulsion module. *Acta Astronautica*, 91:20–33, 2013.

[19] Quan Wen, Liwei Yang, Shanghong Zhao, Yingwu Fang, and Yi Wang. Removing small scale space debris by using a hybrid ground and space based laser system. *Optik*, 141:105–113, 2017.

[20] S. Kitamura. Large space debris reorbiter using ion beam irradiation. *Paper presented at 61st International Astronautical Congress, Prague*, 2010.

[21] Claudio Bombardelli and Jesus Pelaez. Ion beam shepherd for contactless space debris removal. *Journal of Guidance, Control, and Dynamics*, 34(3):916–920, 2011.

[22] R.V. Akhmetzhanov, A.V. Bogatyy, V.G. Petukhov, G.A. Popov, and S.A. Khartov. Radio-frequency ion thruster application for the low-orbit small SC motion control. *Advances in the Astronautical Sciences*, 161:979–989, 2018.

[23] G.G. Raikunov, V.A. Komkov, V.M. Mel'nikov, and B.N. Kharlov. Centrifugal frameless large space structures. *Moscow, ANO "Fizmatlit" Publ*, 2009 [in Russian].

[24] Y. Tsuda, O. Mori, R. Funase, H. Sawada, T. Yamamoto, T. Saiki, T. Endo, and J. Kawaguchi. Flight status of IKAROS deep space solar sail demonstrator. *Acta Astronautica*, 69:833–840, 2011.

[25] *ISO 24113:2019 Space systems – Space debris mitigation requirements*, volume 3. Technical Committee: ISO/TC 20/SC 14 Space systems and operations, 2019.

[26] Anne Dorothy Marinan. *From CubeSats to constellations: systems design and performance analysis*. PhD thesis, Massachusetts Institute of Technology, 2013.

[27] V.G. Petukhov, W.S. Wook, and M.S. Konstantinov. Simultaneous optimization of the low-thrust trajectory and the main design parameters of the spacecraft. *Advances in the Astronautical Sciences*, 161:639–655, 2018.

[28] Brent William Barbee, Salvatore Alfano, Elfego Pinon, Kenn Gold, and David Gaylor. Design of spacecraft missions to remove multiple orbital debris objects. In *35th Annual AAS Guidance and Control Conference, Colorado*, pages 1–14. IEEE, 2012.

[29] Jing Yu, Xiao-qian Chen, and Li-hu Chen. Optimal planning of LEO active debris removal based on hybrid optimal control theory. *Advances in Space Research*, 55(11):2628–2640, 2015.

[30] Max Cerf. Multiple space debris collecting mission—debris selection and trajectory optimization. *Journal of Optimization Theory and Applications*, 156(3):761–796, 2013.

[31] A.A. Baranov, D.A. Grishko, V.V. Medvedevskikh, and V.V. Lapshin. Solution of the flyby problem for large space debris at sun-synchronous orbits. *Cosmic Research*, 54(3):229–236, 2016.

[32] Hironori Sahara. Evaluation of a satellite constellation for active debris removal. *Acta Astronautica*, 105(1):136–144, 2014.

[33] V. Braun, A. Lüpken, S. Flegel, J. Gelhaus, M. Möckel, C. Kebschull, C. Wiedemann, and P. Vörsmann. Active debris removal of multiple priority targets. *Advances in Space Research*, 51(9):1638–1648, 2013.

[34] P. Tadini, U. Tancredi, M. Grassi, L. Anselmo, C. Pardini, A. Francesconi, F. Branz, F. Maggi, M. Lavagna, and L.T. DeLuca. Active debris multi-removal mission concept based on hybrid propulsion. *Acta Astronautica*, 103:26–35, 2014.

[35] A.A. Baranov, D.A. Grishko, and Y.N. Razoumny. Large-size space debris flyby in low earth orbits. *Cosmic Research*, 55(5):361–370, 2017.

[36] A.A. Baranov, D.A. Grishko, Y.N. Razoumny, and L. Jun. Flyby of large-size space debris objects and their transition to the disposal orbits in LEO. *Advances in Space Research*, 59(12):3011–3022, 2017.

[37] K.T. Alfriend, D. Lee, and N.G. Creamer. Optimal servicing of geosynchronous satellites. *Journal of Guidance, Control, and Dynamics*, 29(1):203–206, 2006.

[38] Xiao-qian Chen and Jing Yu. Optimal mission planning of GEO on-orbit refueling in mixed strategy. *Acta Astronautica*, 133:63–72, 2017.

[39] K. Daneshjou, A.A. Mohammadi-Dehabadi, and Majid Bakhtiari. Mission planning for on-orbit servicing through multiple servicing satellites: a new approach. *Advances in Space Research*, 60(6):1148–1162, 2017.

[40] Yu Jing, Xiao-qian Chen, and Li-hu Chen. Biobjective planning of GEO debris removal mission with multiple servicing spacecrafts. *Acta Astronautica*, 105(1):311–320, 2014.

[41] Jing Yu, Xiao-qian Chen, Li-hu Chen, and Dong Hao. Optimal scheduling of GEO debris removing based on hybrid optimal control theory. *Acta Astronautica*, 93:400–409, 2014.

[42] A.A. Baranov, D.A. Grishko, O.I. Khukhrina, and Danhe Chen. Optimal transfer schemes between space debris objects in geostationary orbit. *Acta Astronautica*, 169:23–31, 2020.

[43] V. Zelentsov, G. Shcheglov, V. Mayorova, and T. Biushkina. Spacecrafts service operations as a solution for space debris problem. *International Journal of Mechanical Engineering and Technology*, 9:1503–1518, 2018.

[44] Deutsches Zentrum fur Luft– und Raumfahrt. "ROTEX (1988–1993), Robot Technology Experiment on Spacelab D2-Mission", Closed Space Robotics Missions, DLR Institute of Robotics and Mechatronics (accessed December 2019). http://www.dlr.de/rm/en/desktopdefault.aspx/tabid3827/5969_read8744/.

[45] B. Brunner, G. Hirzinger, K. Landzettel, and J. Heindl. Multisensory shared autonomy and tele-sensor-programming – Key issues in the space robot technology experiment ROTEX. In *Proceedings of 1993 IEEE/RSJ International Conference on Intelligent Robots and Systems (IROS '93)*, volume 3, pages 2123–2139, 1993.

[46] Deutsches Zentrum fur Luft– und Raumfahrt. "the ESS (Experimental Servicing Satellite)", Closed Space Robotics Missions, DLR Institute of Robotics and Mechatronics (accessed December 2019). https://www.dlr.de/rm/en/desktopdefault.aspx/tabid-3827/5969_read-8750/ .

[47] M.H. Kaplan. Space debris realities and removal. In *SOSTC Improving Space Operations Workshop Spacecraft Collision Avoidance and Co–location. The Johns Hopkins University, Applied Physics Laboratory*, 2010.

[48] J. Starke. ROGER a potential orbital space debris removal system. In *Paper presented at 38 COSPAR Scientific Assembly*, 2010.

[49] European Space Agency. "Automation & Robotics" (accessed December 2019). http://www.esa.int/Enabling_Support/Space_Engineering_Technology/Automation_and_Robotics/Automation_Robotics.

[50] Thomas Debus and Sean Dougherty. Overview and performance of the front-end robotics enabling near-term demonstration (FREND) robotic arm. In *AIAA Infotech@Aerospace Conference*, 2009.

[51] B. E. Kelm, J. A. Angielski, S. T. Butcher, N. G. Creamer, K. A. Harris, C. G. Henshaw, J. A. Lennon, W. E. Purdy, F. A. Tasker, and W. S. Vincent. FREND: pushing the envelope of space robotics. Technical report, Naval Research Lab Washington DC, 2008.

[52] NASA Satellite Servicing Projects Division. The robotic servicing arm. https://sspd.gsfc.nasa.gov/robotic_servicing_arm.html.

[53] National Aeronautics and Space Administration. "Satellite Servicing Projects Division (SSPD)" (accessed December 2019). "https://sspd.gsfc.nasa.gov/".

[54] National Aeronautics and Goddard Space Flight Center Space Administration. "On-Orbit Satellite Servicing Study", Project Report, October 2010. https://sspd.gsfc.nasa.gov/images/nasa_satellite%20servicing_project_report_0511.pdf.

[55] National Aeronautics and Space Administration. "Restore-L Proving Satellite Servicing". https://www.nasa.gov/sites/default/files/atoms/files/restore_l_factsheet_092717.pdf, 2016.

[56] G. Roesler. "Robotic Servicing of Geosynchronous Satellites (RSGS) Program Overview", DARPA. `http://images.spaceref.com/fiso/2016/061516_roesler_darpa/Roesler_6-15-16.pdf`, 2016.

[57] SSL corporation. "Space Infrastructure Services" (accessed December 2019). `http://spaceinfrastructureservices.com`.

[58] B. L. Benedict. Rationale for need of in-orbit servicing capabilities for GEO spacecraft. In *AIAA SPACE 2013 Conference and Exposition*, page 5444, 2013.

[59] Intelsat General Corporation. "Intelsat Media Backgrounder on Satellite Refueling & Service". `https://www.intelsatgeneral.com/wp-content/uploads/files/Intelsat%20Media%20Backgrounder%20on%20Refueling.pdf`, 2019.

[60] MDA corporation. `https://mdacorporation.com`, 2018.

[61] Shin-Ichiro Nishida and Naohiko Kikuchi. A scenario and technologies for space debris removal. In *12th International Symposium on Artificial Intelligence, Robotics and Automation in Space, i-SAIRAS*, volume 14, 2014.

[62] Les Johnson, Mark Whorton, Andy Heaton, Robin Pinson, Greg Laue, and Charles Adams. Nanosail-D: A solar sail demonstration mission. *Acta Astronautica*, 68(5-6):571–575, 2011.

[63] Andrew F. Heaton, Brent F. Faller, and Chelsea K. Katan. Nanosail-D orbital and attitude dynamics. In *Advances in Solar Sailing*, pages 95–113. Springer, 2014.

[64] Chris Biddy and Tomas Svitek. LightSail-1 solar sail design and qualification. In *Proceedings of the 41st Aerospace Mechanisms Symposium*, pages 451–463. Jet Propulsion Lab., National Aeronautics and Space Administration. Pasadena, CA, 2012.

[65] S. Stirone. Lightsail-2 Unfurls, Next Step Toward Space Travel by Solar Sail – deployed LightSail 2, aiming to further demonstrate the potential of the technology for space propulsion. `https://www.nytimes.com/2019/07/23/science/lightsail-solar-sail.html`, July 2019.

[66] V. Lappas, N. Adeli, L. Visagie, J. Fernandez, T. Theodorou, W. Steyn, and M. Perren. CubeSail: A low cost CubeSat based solar sail demonstration mission. *Advances in Space Research*, 48(11):1890–1901, 2011.

CHAPTER 5

Space Debris Mitigation Based on Commercial Off-the-Shelf Technologies

Shinichi Kimura

Tokyo University of Science, Tokyo, Japan

CONTENTS

T HERE is concern that the amount of space debris is increasing continuously because of how it self-regenerates through collision. Therefore, we need to establish mitigation technologies, collectively called active debris removal (ADR), to remove existing space debris and ensure the sustainable utilization of near-Earth orbits. The capturing and rendezvous processes are among the most important and difficult processes involved in ADR. In addition, the high cost of ADR systems is a major obstacle to their implementation. In this section, we discuss our low-cost intelligent guidance and navigation system that will enable reliable mitigation of space debris.

5.1   INTRODUCTION

Problems associated with space debris are becoming increasingly serious as space development becomes more intense [1, 2]. There is concern that the amount of space debris is increasing continuously via self-regeneration through collision, which is known as the Kessler syndrome. Therefore, we must not only control the generation of future space debris, but also establish mitigation technologies to remove existing space debris to ensure the sustainable utilization of near-Earth orbits. Debris removal technologies are collectively called active debris removal (ADR).

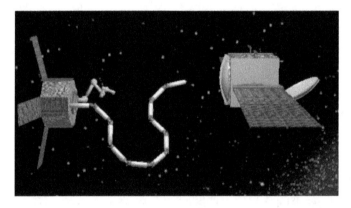

Figure 5.1: Concept of Orbital Maintenance System (OMS)

Concepts to remove space debris using space robots emerged as part of the on-orbit satellite servicing concepts, such as the Geostationary Service Vehicle proposed by ESA to utilize orbits efficiently by repairing and refuelling geostationary satellites [3]. The Orbital Maintenance System (OMS) is the concept to remove used and/or failed satellites to maintain clean orbits [4, 5] (Fig. 5.1). This idea includes a concept called OMS Lights, which is a stepwise demonstration that effectively utilizes small satellites. By stepwise demonstration, we can reduce the cost to develop the technologies for on-orbit servicing and demonstrated technologies that are expected to generate applications (Fig. 5.2). To realize on-orbit satellite maintenance, improvements in various technologies are required, but the development costs are expected to decrease because of their substantial role in space development. In the early 2000s,

Kimura proposed a demonstration mission of technologies that rendezvous with and fly around the uncooperative target called SmartSat-1 [6–10]. This concept includes small equipment to support rendezvous and capturing, called a rescue package. It is similar to the satellite end-of-life support concept of Astroscale [11–14].

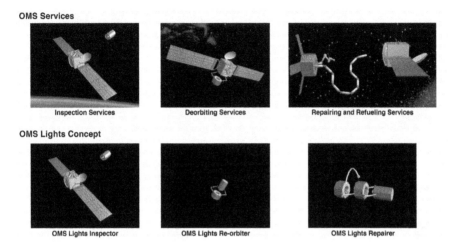

Figure 5.2: Concept of stepwise demonstration of OMS technologies using small satellites

However, the SmartSat-1 could not obtain a flight opportunity, partly because the situational awareness of space debris was still premature and the problem of how to pay for the mitigation of space debris remained. The space debris problem is considered a highly public issue. Therefore, there was not enough impetus to fund space debris mitigation. We can learn a lesson from this result. In order to realize space debris mitigation, we need both public awareness and a significant reduction in development costs. Space debris can be considered the industrial waste of space development, in that only a few organizations can afford the large cost.

Many activities are emerging for space debris mitigation, such as those of Astroscale [11–14] and Kawasaki Heavy Industries (KHI) [15, 16] in the private sector. In addition, the CleanSpace One project of the EPFL Space Center (eSpace) is a unique approach in space debris mitigation [17]. Such activities demonstrate the importance of space debris mitigation.

Even with such activities, reductions in cost are very important. In a sense, the cost is essential because space debris removal is driven by the private sector. One of the most important factors in reducing the cost of space systems is the utilization of terrestrial technologies, especially commercial off-the-shelf (COTS) devices. Because the market for space devices is still much smaller than for terrestrial devices, the space devices are very costly and their functions are limited. The space environment is different from the terrestrial environment, but the devices can be qualified using simulations. If we can utilize COTS devices with proper qualifications, we can develop space equipment with a high functionality and a very low cost.

As Fig. 5.3 shows, a typical ADR process consists of several steps. Once the ADR satellite is launched into an orbit near the target debris, the ADR satellite needs to gain access to the target debris. Next, the ADR satellite captures the target debris and attaches deorbiting equipment to the debris. Finally, the target debris is deorbited by the deorbiting equipment. The capturing and rendezvous processes are among the most important and difficult processes involved in ADR. These processes are critical because an unexpected collision with the target may cause not only the mission to fail, but also additional generation of space debris.

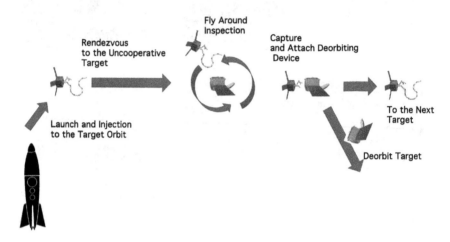

Figure 5.3: Sequence of Active Debris Removal (ADR)

The orbit of an ADR satellite can be approximately controlled using orbit information, but the ADR satellite needs to identify the target debris orbit precisely to achieve a fine-tuned and safe rendezvous with the target debris. This is because the orbit information has an uncertainty associated with it. The rendezvous process can be partly teleoperated and supported by ground operations, but it needs to be performed by the ADR satellite autonomously because the communication link between the ADR satellite and ground station is limited. In the case of a low-Earth orbit in which an ADR typically operates, one communication path is only 10 minutes per ground station, and only five or six paths are available per day. In orbit, an ADR satellite also needs to adapt to various situations and be able to function safely in various situations. To autonomously access the target debris safely and reliably, an ADR satellite needs to have "eyes" to find the target debris and a "brain" to control the rendezvous process autonomously [4–10, 18–21].

After an ADR satellite accesses a target, it needs to capture the target debris using a space robot to perform actions such as attaching deorbiting equipment to the target debris. Capturing target debris in free-flying situations is very difficult for an ADR robot because unanticipated contact can produce unexpected motion. Thus, the robot needs to deduce the relative motion required to reach the target and capture it securely with a simple action. In addition, it is very difficult to teleoperate such a process directly because of the communication time delay involved. Therefore, we need highly intelligent robotic hands to capture the target debris.

Space debris mitigation is affected by strong cost pressures. Therefore, high-performance hand, eye, and brain activities need to be conducted at extremely low costs to perform space debris mitigation. In the remainder of this section, we describe a few achievements in the development of key technologies for space debris mitigation, and we discuss other challenges facing ADR.

5.2 GUIDANCE AND NAVIGATION TECHNOLOGIES FOR RENDEZVOUS WITH TARGET DEBRIS

As Fig. 5.4 shows, the process by which the ADR satellite accesses the target debris can be divided into three phases [4–10, 18–21]. First, the ADR satellite is launched into the same orbit as the target debris. Then, based on orbit information that is estimated using a radar network, the ADR satellite accesses the target debris by means of commands sent from the ground station. This process is called the far-distance phase. Because the uncertainty in the orbit estimation is approximately a few kilometres, the ADR satellite may be a few kilometres from the target debris at the end of the far-distance phase. At such a distance, when the ADR satellite acquires a visual image of the target debris, it can be recognized only as a light dot. Active sensors such as radar cannot be used to recognize the target debris at this distance because of the high energy consumption that they would require.

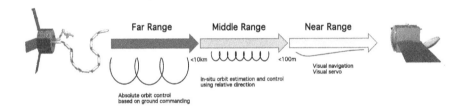

Figure 5.4: Rendezvous process

When the ADR satellite comes within a few tens of meters of the target debris, it can recognize the target debris as a shaped object and can recognize its motion based on a feature tracking technique that uses visual images of the target debris. Active sensors such as radar can also be utilized at such distances. This is called the near-distance phase or final approach phase. In this phase, robot vision technologies such as visual servo or feature-based target recognition can also be used.

The most difficult and most important phase between the far-distance phase and the near-distance phase is the phase corresponding to distances from a few kilometres to a few tens of meters from the target debris. This is called the middle-distance phase. In this phase, active sensors cannot be utilized because they are costly, heavy, and consume large amounts of energy, all of which are significant obstacles to the development of low-cost ADR satellites. We have found that an orbit estimation technique that uses visual images is valuable in overcoming such obstacles and achieving complete rendezvous manoeuvres with target debris. Using this estimation technique, the target debris orbit can be estimated from the ADR orbit, and the target direction sequence can be determined from visual images.

Based on the considerations discussed above, we focused on the development of a low-cost intelligent space camera for use in a guidance and navigation system for space debris mitigation. A low-cost intelligent space camera offers the following advantages:

1. Orbit estimation using direction information acquired from visual images may make it possible to achieve a rendezvous in the middle-distance phase.

2. During critical operations such as a rendezvous with the space debris, visual information is indispensable not only to achieve safe and reliable operation, but also to understand the situation in which the spacecraft is operating.

3. In comparison to active sensors, an intelligent camera has lower cost, weight, size, and energy requirements.

Precise positioning and orientation information are needed to capture target debris. Therefore, the ADR satellite should have sufficiently high intelligence to acquire the required information and to be able to control its accessing strategy autonomously. In addition, its image and information processing capabilities need to be highly flexible because the situations that ADR spacecraft may encounter are complicated. However, ADR spacecraft face strong cost pressures because of their highly complex functionality within a space system. The requirement to keep costs low affects not only the development costs of the spacecraft itself, but also the costs of various resources related to size, weight, and power. Therefore, we need to establish highly intelligent and flexible guidance and navigation camera systems within severe cost and resource constraints.

5.3 UTILIZATION OF COMMERCIAL OFF-THE-SHELF (COTS) TECHNOLOGIES FOR INTELLIGENT SPACE CAMERA

It is difficult to develop low-cost intelligent space cameras for space debris mitigation activities using space-specialized devices because they are high in cost and poor in performance. The key to developing high-performance, low-cost guidance and navigation systems is effective utilization of commercial off-the-shelf devices. To utilize COTS devices in space, we need to be aware of the major differences between terrestrial utilization and orbital utilization. When we understand this difference and apply appropriate qualification processes, the implementation of orbit utilization is less challenging. The major difference between terrestrial utilization and orbital utilization can be summarized as follows: if the two aspects described in the following subsections are carefully considered, COTS devices can be used in Earth orbit.

5.3.1   Thermal vacuum condition

There is no atmosphere in space. On Earth, air is crucial to removing heat generated by electrical devices. However, in space, this heat is removed by conduction through a circuit board or by radiation. Thermal conditions vary significantly around the Earth's orbit. Static heat control and dynamic changes in thermal conditions affect the soldering of the circuit boards used for conduction. Therefore, circuit boards and other relevant devices need to be tested under thermal vacuum conditions. With regard to soldering, another phenomenon causes problems in a vacuum. If lead-free solder is used in circuit boards that are launched into space, needle-like protuberances called "whiskers" emerge from the solder. For orbital utilization, this type of soldering is preferred [21].

In addition to static thermal vacuum conditions, in some cases we need to address the problem of thermal shock. The temperature can change suddenly when a spacecraft experiences night conditions and then suddenly experiences daylight conditions. A thermal shock test facility can be effective in quantifying the effects of thermal shock (Fig. 5.5). Such a facility consists of two chambers: a heating chamber and a cooling chamber. A small elevator is located between the two chambers. A sudden temperature change is induced when the elevator quickly moves the device from one chamber to the other. In addition to measuring resistance to thermal shock, the thermal shock test facility can be used for ageing acceleration tests.

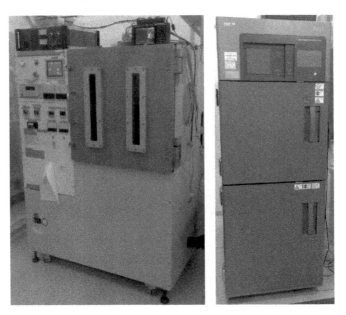

Figure 5.5: Thermal vacuum and thermal shock test facilities

5.3.2 Radiation condition

Although the Earth's atmosphere shields us from the sun's radiation, the radiation and particles emitted by the sun affect electronic devices in space. This radiation effect is a major difference between the terrestrial and orbital environments and can be divided into two categories: the total ionizing dose (TID) and the single-event effect (SEE). The TID is an irreversible degradation of a device caused by exposure to large amounts of radiation. This effect depends heavily on the material and fabrication process of the device and indicates the possible lifetime of devices in orbit. The TID can be measured by a gamma radiation test. The validation and screening of devices with the TID are considered effective protective measures against harmful radiation the devices will experience [19–24].

The SEE is a probabilistic effect caused by radiated particles. It can lead to single-event upsets (SEUs), which cause bit inversion of logical devices and memories. The effect of an SEU is based on characteristics measured by a particle radiation test. If necessary, software-based countermeasures can be adopted, such as the installation of an error correction code, to protect against SEUs. Simple intermittent resetting of the central processing unit (CPU) is also a countermeasure against SEUs for simple applications. The SEE can also cause single-event latch-ups (SELs), which cause tentative overcurrent conditions. In general, an SEL can be addressed by a swift power reset of the relevant devices.

The environment in space can be simulated using test facilities such as ultra-vacuum chambers that simulate the ultra-vacuum conditions in orbit and radiation facilities that simulate radiation conditions (see Fig. 5.6). The performance of space-based architecture can be tested on the ground using such simulations.

Figure 5.6: Radiation test facilities. Cobalt 60 Test Facility of Laboratory for Advanced Nuclear Energy, Tokyo Institute of Technology (Left) and Cyclotron Test Facility of Quantum Medical Science Directorate, National Institute of Quantum and Radiological Science and Technology (Right)

Software technology can also be used to overcome the effects of the space environment. For example, to address the bit-alternating effect of an SEU, error correction technologies can be used for memory devices. Similarly, if the electric current in a power line is monitored and devices are reset as soon as overcurrent is detected, the effects of an SEL can be mitigated. Measures such as these make it possible for COTS devices to be utilized in orbit.

5.4 VISUAL GUIDANCE AND NAVIGATION SYSTEM FOR SPACE DEBRIS REMOVAL USING COTS TECHNOLOGIES

We need to develop intelligent robotic eyes and a brain to successfully navigate the ADR satellite to the target debris at a very low cost. To overcome the difficulties involved, we have developed a compact intelligent guidance and navigation camera system for space debris mitigation [21–28] (Fig. 5.7). The key aspects of this camera system are the following: (1) effective qualification and utilization of COTS technologies, (2) effective utilization of field-programmable gate array (FPGA) technologies for first-image processing, and (3) flexible software implementation using the Linux operating system. The image-processing unit used in this system has already been proven in space on various missions, such as IKAROS, Hayabusa-2, and KITE [20, 29–31]. Plans are in place to utilize and demonstrate the camera system in the guidance and navigation of the on-orbit missions of Astroscale and KHI.

In this section, we describe the low-cost intelligent guidance and navigation system that has been designed to enable reliable mitigation of space debris.

Figure 5.7: Guidance and navigation camera system

5.4.1 System architecture

The guidance and navigation camera system consists of camera head units and a processing unit. These units are interconnected using a high-speed serial interface. By separating the compact camera head unit from the processing unit, we can easily place the camera head units to avoid mechanical or field-of-view (FOV) conflicts with other pieces of equipment. One processing unit can support a maximum of two camera head units, and we can select a few FOV options for the lens of each camera unit. One idea is that we can utilize two cameras to obtain stereo images using the same FOV lens. Another idea is that we can utilize wide and narrow lens sets in a manner similar to that used with KITE-CAM. Wide FOV cameras are effective for searching and finding target debris and narrow FOV cameras are useful in obtaining precise and high-resolution information on target debris. The processing unit consists of an FPGA board and interface board, which interconnects the FPGA board with the camera head units. As Fig. 5.7 shows, the camera system can be assembled as a printed circuit board structure and as a component structure. Therefore, it can be implemented as a part of a component to minimize its size and weight requirements.

5.4.2   Electrical system

The camera system has two image lens systems, as described above, and employs an interface (IF) board to control these systems using a processing system. The wiring is reduced through the use of a high-speed serializer and a deserializer. The IF board alternates between the two image lens systems via a bus switch. Figure 5.8 shows a diagram of the system. This architecture has already been proven in space during the KITE mission.

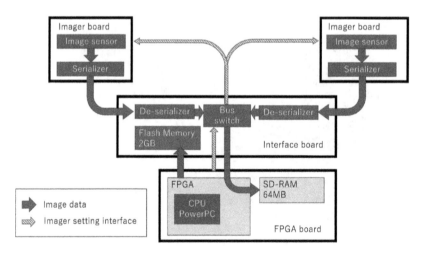

Figure 5.8: System diagram. SDRAM: Synchronous Dynamic Random Access Memory

With respect to the image sensors of the camera heads, there are a few options for resolution, frequency, and radiation tolerance, depending on the requirements. The OV9630 features one-megapixel middle-range resolution, and exhibits high radiation tolerance and reliability and is utilized in many space applications. The highest-resolution version produces a five-megapixel image that is radiation qualified and will be utilized in the Astroscale mission. We also have a high-sensitivity and high-frequency version as an additional option for these camera heads.

Figure 5.9 shows the FPGA board, for which we installed a Xilinx FPGA Virtex-2 Pro as the main processor. The board includes a processor and a memory unit that have been proven in various space missions, including IKAROS, Hayabusa-2, and KITE. This FPGA board can connect to the IF board in a 5-mm stack.

In many cases, the visual guidance and navigation system needs to provide raw image data to validate image-processing functions. The camera system uses 2 GB of NAND flash memory to store raw images during real-time image processing. The stored images can then be downloaded on command.

Figure 5.9: FPGA board

5.4.3   FPGA for image preprocessing and interface

In guidance and navigation applications, it is important to maintain real-time processing during the catering time interval for purposes such as target recognition and/or tracking. To speed up the image processing function, the camera system can utilize an FPGA preprocessing function. The processing unit utilizes FPGA for the image acquisition process, which means that the preprocessing function can be included in the image acquisition process. We implemented a projection function in the image acquisition process to identify the target's position in far-distance situations, but the preprocessing function can be modified depending on the application.

As the CPU bus structure is implemented in the FPGA, the interface function of the CPU can also be customized by, for example, increasing the interface channels according to the application. By utilizing such customizable features of the interface function, we are able to implement the interface to the transmitter successfully. This interface allows us to transmit the captured images directly to the ground. Such a capability is very useful in monitoring the mission process.

5.4.4 Software implementation and development environment

The Linux (kernel 2.4) OS was employed in the image processor. Linux is very flexible and can utilize various free open-source applications, which makes it possible to shorten the development period. In addition, using a memory management unit in combination with Linux, both of which are installed in the CPU, can ensure powerful memory protection and prevent important data from suddenly being destroyed.

5.4.5 Achievements in space robotics

We have considerable experience with on-orbit hardware technologies for space robots. A few robotic arms are already operating on the International Space Station, not only on experimental missions such as REXJ (Fig. 5.10), but also as infrastructure components, such as SSRMS and JEMRMS. Japan has developed a few robotic arms and has operated them in orbit since the first space robot experiment on the space shuttle, known as the Manipulator Flight Demonstration [32] (Fig. 5.11).

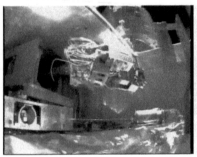

Figure 5.10: REXJ mission

Figure 5.11: Manipulator flight demonstration

In 1998, Japan successfully launched the world's first space robotic satellite, known as the Engineering Test Satellite VII (ETS-VII: Fig. 5.12, and has operated space robots in unmanned situations [33–36]. These missions confirmed that shared intelligence technologies, such as compliance control technologies, are very effective in compensating for uncertainty in remote unmanned situations. In addition, a few novel teleoperation technologies, including multi-modal interfaces, have been shown to be effective at operating securely under critical operation conditions. The world's first unmanned capture and berthing experiment was successfully conducted using ETS-VII. Therefore, certain techniques crucial to space debris mitigation have already been demonstrated to be successful.

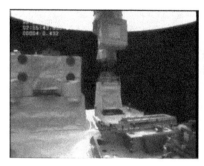

Figure 5.12: Engineering Test Satellite VII (ETS-VII)

Even though some technical obstacles have already been overcome, at least two aspects remain to be addressed in the successful implementation of ADR satellites for space debris mitigation. The first point is how to capture an uncooperative target. Space debris is not designed to be captured. Therefore, there is no capturing point such as a fixture or a handle. It is very difficult to capture a target that is not designed to be captured. A few trials have been conducted by various research groups to address this problem. Their approaches include the utilization of a capturing thruster nozzle [15] and a net or harpoon [37]. However, no one fail-proof method has yet been identified to safely and reliably capture space debris. The idea associated with Astroscale to install effective contact points for future satellite end-of-life treatment is unique and effective, but we need to know how to capture current space debris [11–14].The second point is the cost of the robotic arm. The robotic arm is still too expensive to be implemented for space debris removal. Mechanical systems designed with high precision and reliability are expensive. It may be possible to reduce the cost by considering optimization functions and utilization of COTS technologies for the controller in a manner similar to that for the guidance and navigation system. These two aspects remain to be addressed to optimize the capture of space debris, and therefore we need to seek ways to accomplish active debris removal more effectively.

To address the pressure to control costs and increase reliability using the fault-adaptive concept, we are studying decentralized module-type robot systems that adapt to partial failure autonomously [38–40] (Fig. 5.13). Robotic manipulators can be constructed in the same manner as modularized robots using the aforementioned system, and the manipulators can perform their tasks even though a few modules may fail. Based on this concept, we can reduce the reliability requirement and reduce the cost of robotic manipulators.

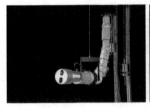

Figure 5.13: Module-type manipulator

5.5 CONCLUSION

To solve the space debris problem and maintain sustainable and safe Earth orbits, it is important to reduce the future production of space debris by actively removing the existing, in-situ debris, given that it can lead to additional collisions and break-ups. Various technologies, such as in-situ identification of space debris, rendezvous and homing, and robotics, need to be improved to actively remove space debris. Simultaneously, we need to maintain the costs incurred in the implementation of these technologies at low levels, although they require increased autonomy and performance. In this chapter, we discussed a few ideas to harmonize the technological requirements and alleviate the pressure for their realization. Nevertheless, it is believed that these may not be sufficient because the required technologies are wide-spreading. Therefore, researchers and engineers, particularly the younger generation, should be encouraged to tackle technologies for active debris removal. Sustainable and safe Earth orbits partly depend on noble and unique ideas that aim to mitigate these technical issues.

ACRONYM

ADR Active Debris Removal

COTS Commercial Off-The-Shelf

CPU Central Processing Unit

eSpace EPFL Space Center

ETS-VII Engineering Test Satellite VII

EPFL École polytechnique fédérale de Lausanne

FPGA Field-Programmable Gate Array

FOV Field-Of-View

IF InterFace

IKAROS Interplanetary Kite-craft Accelerated by Radiation Of the Sun

JAXA Japan Aerospace Exploration Agency

JEM Japanese Experiment Module

JEMRMS JEM Remote Manipulator System

KITE Kounotori Integrated Tether Experiments

KHI Kawasaki Heavy Industries, Ltd.

OMS Orbital Maintenance System

OS Operating System

REXJ Robot Experiment on JEM

SEE Single-Event Effect

SEL Single-Event Latch-up

SEU Single-Event Upsets

SSRMS Space Station Remote Manipulator System

TID Total Ionizing Dose

GLOSSARY

Active Debris Removal: A service or system to deorbit and remove orbital debris actively

Commercial Off-The-Shelf: Products that are ready-made and available for sale to the general public.

Total Ionizing Dose: An irreversible degradation of a device caused by exposure to large amounts of radiation

Single-Event Effect: A probabilistic effect caused by radiated particles

Single-Event Upsets: A SEE that causes bit inversion of logical devices and memories

Single-Event Latch-up: A SEE that causes tentative overcurrent conditions

Field-Programmable Gate Array: An integrated circuit designed to be configured by a customer or a designer after manufacturing

FURTHER READING

Kimura S., Asakura Y., Doi H. and Nakamura M. (2017). *Document-based Programming System for Seamless Linking of Satellite Onboard Software and Ground Operating System*. Journal of Robotics and Mechatronics, 29(5), pp. 801-807.

Narumi T., Tsukamoto D. and Kimura S. (2016). *Robust on-orbit optical position determination of non-cooperative spacecraft*. In 26th AAS/AIAA Space Flight Mechanics Meeting, 2016 (pp. 3251-3263).

Kimura S., Narumi T., Aoyanagi Y., and Nakasuka S., (2015). *Optical Space Equipments Using Commercial Off-the-Shelf Devices*. Optical Payloads for Space Missions (Shen-En Qian Ed. John Wiley & Sons ISBN: 978-1-118-94514-8).

Kobayashi S., Takisawa J., Nakasuka S., and Kimura S. (2014). *Software Development Framework for Small Satellite On-board Computers* ransactions of the Japan Society for Aeronautical and Space Sciences, Aerospace Technology Japan. 12(ists29), Tf_1-Tf_6

Kimura S., Takeuchi M., Harima K., Fukase Y., Sato H., Yoshida T., Miyasaka A., Noda H., Sunakawa K., and Homma M. (2004). *Visual Analysis in a Deployable Antenna Experiment*. IEEE Transactions on Aerospace and Electronic Systems, 40(1), pp. 247-258.

Kimura S., Tsuchiya S., Takegai T., and Nishida S., (2001). *Fault Adaptive Kinematic Control Using Multiprocessor System and its Verification Using a Hyper-Redundant Manipulator*. Journal of Robotics and Mechatronics, 13(5), pp. 540-547.

Kimura S., Tsuchiya S., Nagai Y., Nakamura K., Satoh K., Morikawa H., and Takanashi N., (2000). *Teleoperation Techniques for Assembling an Antenna by Using Space Robots - Experiments on Engineering Test Satellite VII*. Journal of Robotics and Mechatronics, 12(4), pp. 394-401.

Kimura S., Yano M., and Shimizu H., (1994). *A Self-Organizing Model of Walking Patterns of Insect. II. The Loading Effect and Leg Amputation*. Biological Cybernetics, 70(6), pp. 505-512.

Kimura S., Yano M., and Shimizu H., (1993). *A Self-Organizing Model of Walking Patterns of Insects*. Biological Cybernetics, 69(3), pp. 183-193.

Bibliography

[1] E.L. Christiansen, D.M. Lear, and J.L. Hyde. Recent impact damage observed on international space station. *Orbital Debris Quarterly News*, 18(4):3–4, 2014.

[2] J.C. Liou. An update on leo environment remediation with active debris removal. *The NASA Orbital Debris Program Office, Quarterly News*, 15(2):4–6, 2011.

[3] G. Visentin and D.L. Brown. Robotics for geostationary satellite servicing. *Robotics and Autonomous Systems*, 23(1-2):45–51, 1998.

[4] S. Kimura, S. Tsuchiya, K. Araki, Y. Suzuki, and R. Suzuki. OMS for NeLS a concept for a robot-assisted service for removing satellites from a LEO constellation. In *International Astronautical Federation-51st International Astronautical Congress 2000 (IAC 2000), Rio de Janeiro, Brazil, IAA-00-IAA.6.6.02*, 2000.

[5] S. Kimura, H. Mineno, H. Yamamoto, Y. Nagai, H. Kamimura, S. Kawamoto, F. Terui, S. Nishida, S. Nakasuka, and S. Ukawa. Preliminary experiments on technologies for satellite orbital maintenance using Micro-LabSat 1. *Advanced Robotics*, 18(2):117–138, 2004.

[6] S. Kimura, H. Mineno, H. Yamamoto, Y. Nagai, H. Kamimura, S. Kawamoto, F. Terui, S. Nishida, S. Nakasuka, and S. Ukawa. Preliminary experiments on image processing for satellite orbital maintenance. In *Proceedings of the 7th International Symposium on Artificial Intelligence, Robotics and Automation in Space (i-SAIRAS 2003), Nara, Japan, AS02*, 2003.

[7] S. Kimura, Y. Nagai, H. Yamamoto, T. Kashitani, K. Masuda, and N. Abe. Experimental concept on technologies for in-orbit maintenance using a small twin-sat. In *International Astronautical Federation-55th International Astronautical Congress 2004 (IAC 2004), Vancouver, Canada, IAC-04-U.4.05*, 2004.

[8] S. Kimura, N. Nishinaga, M. Akioka, N. Abe, K. Masuda, and S. Nakamura. SmartSat-1: On orbit experiment plan using mini-satellite. In *Workshop for Space, Aeronautical and Navigational Electronics 2005 (WSANE2005), Daejeon, Korea*, pages 137–142, 2005.

[9] S. Kimura, Y. Nagai, H. Yamamoto, H. Kozawa, S. Sugimoto, S. Nakamura, K. Masuda, and N. Abe. Rendezvous experiments on SmartSat-1. In *2nd IEEE International Conference on Space Mission Challenges for Information Technology (SMC-IT'06)*, pages 374–379. IEEE, 2006.

[10] S. Kimura, Y. Nagai, H. Yamamoto, H. Kozawa, S. Sugimoto, K. Masuda, and N. Abe. Possibility of small satellite in on-orbit servicing. In *25th International Symposium on Space Technology and Science Ishikawa, Japan*, pages 1618–1623, 2006.

[11] M. Okada, A. Okamoto, K. Fujimoto, and M. Ito. Maximizing post mission disposal of mega constellations satellites reaching end of operational lifetime. In *ESA 7th European Conference on Space Debris, ESOC, Germany*, 2017.

[12] C. Blackerby, A. Okamoto, Y. Kobayashi, K. Fujimoto, Y. Seto, S. Fujita, T. Iwai, N. Okada, J. Forshaw, and J. Auburn. The ELSA-d end-of-life debris removal mission: Preparing for launch. In *70th International Astronautical Congress, Washington DC, US, IAC-19,A6,5,2*, 2019.

[13] C. Weeden, C. Blackerby, N. Okada, E. Yamamoto, J. Forshaw, and J. Auburn. Authorization and continuous supervision of Astroscale's de-orbit activities: A review of the regulatory environment for end of life (EOL) and active debris removal (ADR) services. In *70th International Astronautical Congress, Washington DC, US, IAC-19.A6.8*, 2019.

[14] C. Weeden, C. Blackerby, N. Okada, E. Yamamoto, J. Forshaw, and J. Auburn. Industry implementation of the long-term sustainability guidelines: An Astroscale perspective. In *70th International Astronautical Congress, Washington DC, US, IAC-19,E3,4,4*, 2019.

[15] K. Shibasaki, N. Kubota, M. Enomoto, S. Kawamoto, Y. Ohkawa, J. Aoyama, and Y. Katayama. Conceptual study of mechanical and sensing system for debris capturing for PAF. In *30th International Symposium on Space Technology and Science, Kobe, Japan, 2015-k-42*, 2015.

[16] H. Nakamoto, T. Maruyama, and Y. Sugawara. In-orbit demonstration of vision-based navigation and capturing mechanism for active debris removal by microsat "DRUMS". In *32nd International Symposium on Space Technology and Science, Fukui, Japan, 2019-r-24*, 2019.

[17] EPFL Space Center. CleanSpace One. https://www.epfl.ch/research/domains/epfl-space-center/spaceresearch/cleanspaceone_1/, 2019.

[18] S. Kawamoto, Y. Ohkawa, H. Nalanishi, Y. Katayama, H. Kamimura, and S. Kitamura. Active debris removal by a small satellite. In *63rd International Astronautical Congress, Naples, Italy, IAC-12- A6.7.8*, 2012.

[19] T. Kasai, D. Tsuijita, T. Uchiyama, M. Harada, S. Kawamoto, Y. Ohkawa, and K. Inoue. Feasibility study of electrodynamic tether technology demonstration on H-II transfer vehicle. In *6th IAASS International Space Safety Conference, Montreal, Canada*, 2013.

[20] S. Kimura, Y. Horikawa, and Y. Katayama. Quick report on on-board demonstration experiment for autonomous-visual-guidance camera system for space debris removal. *Transactions of the Japan Society for Aeronautical and Space Sciences, Aerospace Technology Japan*, 16(6):561–565, 2018.

[21] S. Kimura and A. Miyasaka. Qualification tests of micro-camera modules for space applications. *Transactions of the Japan Society for Aeronautical and Space Sciences, Aerospace Technology Japan*, 9:15–20, 2011.

[22] J. L: Barth, C.S. Dyer, and E.G. Stassinopoulos. Space, atmospheric, and terrestrial radiation environments. *IEEE Transactions on Nuclear Science*, 50(3):466–482, 2003.

[23] P.E. Dodd and L.W. Massengill. Basic mechanisms and modeling of single-event upset in digital microelectronics. *IEEE Transactions on Nuclear Science*, 50(3):583–601, 2003.

[24] A. Campbell, S. Buchner, E. Petersen, B. Blake, J. Mazur, and C. Dyer. SEU measurements and predictions on MPTB for a large energetic solar particle event. *IEEE Transactions on Nuclear Science*, 49(3):1340–1344, 2002.

[25] D. Falguere, D. Boscher, T. Nuns, S. Duzellier, S. Bourdarie, R. Ecoffet, S. Barde, J. Cueto, C. Alonzo, and C. Hoffman. In-flight observations of the radiation environment and its effects on devices in the SAC-C polar orbit. *IEEE Transactions on Nuclear Science*, 49(6):2782–2787, 2002.

[26] T. Goka, H. Matsumpoto, H. Koshiishi, H. Liu, Y. Kimoto, S. Matsuda, M. Imaizumi, S. Kawakita, O. Anzawa, K. Aoyama, K. Tanioka, S. Ichikawa, T. Sasada, and S. Yamakawa. Space environment & effect measurements from the MDS-1 (Tsubasa) satellite. *Proceedings of International Symposium on Space Technology and Science, Matsue, Japan, ISTS*, 2002.

[27] S. Kimura, Y. Hiroshi, N. Yasufumi, A. Maki, H. Hidekazu, T. Nobuhiro, K. Matsuaki, and Y. Keisuke. Single-event performance of a COTS-Based MPU under flare and non-flare conditions. *IEEE Transactions on Aerospace and Electronic Systems*, 41(2):599–607, 2005.

[28] S. Kimura, Y. Kasuya, and M. Terakura. Breakdown phenomena in SD cards exposed to proton irradiation. *Trans. JSASS Aerospace Tech. Japan*, 12:31–35, 2014.

[29] S. Kimura, A. Miyasaka, R. Funase, H. Sawada, N. Sakamoto, and N. Miyashita. High-performance image acquisition & processing unit fabricated using COTS technologies. *IEEE Aerospace and Electronic Systems Magazine*, 26(3):19–25, 2011.

[30] S. Kimura, A. Miyasaka, R. Funase, H. Sawada, N. Sakamoto, and N. Miyashita. A high-performance image acquisition and processing system for IKAROS fabricated using FPGA and free software technologies. *61st International Astronautical Congress, Prague, CZ, IAC-10.D1.2.10*, 26(3):19–25, 2010.

[31] S. Kimura, M. Terakura, A. Miyasaka, N. Sakamoto, N. Miyashita, R. Funase, and H. Sawada. A high-performance image acquisition and processing unit using FPGA technologies. In *12th International Conference of Pacific-Basin Societies, ISCOPS*, pages 407–414, 2010.

[32] S. Kimura, T. Okyuama, N. Yoshioka, and Y. Wakabayashi. Robot-aided remote inspection experiment on STS-85. *IEEE Transactions on Aerospace and Electronic Systems*, 36(4):1290–1297, 2000.

[33] M. Oda. Experiences and lessons learned from the ETS-VII robot satellite. In *Proceedings 2000 ICRA. Millennium Conference. IEEE International Conference on Robotics and Automation. Symposia Proceedings (Cat. No. 00CH37065)*, volume 1, pages 914–919. IEEE, 2000.

[34] Y. Suzuki, S. Tsuchiya, T. Okuyama, T. Takahashi, Y. Nagai, and S. Kimura. Mechanism for assembling antenna in space. *IEEE Transactions on Aerospace and Electronic Systems*, 37(1):254–265, 2001.

[35] Y. Nagai, S. Tsuchiya, T. Iida, and S. Kimura. Audio feedback system for teleoperation experiments on Engineering Test Satellite VII: System design and assessments using eye mark recorder for capturing task. *IEEE Transactions on Systems Man and Cybernetics Part A*, 32(2):237–247, 2002.

[36] S. Kimura, T. Okuyama, Y. Yamana, Y. Nagai, and H. Morikawa. Teleoperation system for antenna assembly by space robots. In *Telemanipulator and Telepresence Technologies V*, volume 3524, pages 14–23. International Society for Optics and Photonics, 1998.

[37] B. Taylor, G. Aglietti, S. Fellowes, T. Salmon, A. Hall, T. Chabot, and C. Bernal. Removedebris preliminary mission results. In *International Astronautical Congress, Bremen, Germany, IAC-18,A6,5,1*, 2018.

[38] S. Kimura, M. Takahashi, T. Okuyama, S. Tsuchiya, and Y. Suzuki. A fault-tolerant control algorithm having a decentralized autonomous architecture for space hyper redundant manipulators. *IEEE Transactions on Systems, Man, and Cybernetics-Part A: Systems and Humans*, 28(4):521–527, 1998.

[39] S. Kimura and T. Okuyama. Processor performance required for decentralized kinematic control algorithm of module-type hyper-redundant manipulator. *Journal of Robotics and Mechatronics*, 8(5):442–446, 1996.

[40] S. Kimura, M. Yamauchi, and Y. Ozawa. Magnetically jointed module manipulators: New concept for safe intravehicular activity in space vehicles. *IEEE Transactions on Aerospace and Electronic Systems*, 47(3):2247–2253, 2011.

V

Regulation & Legislation

Addressing the Inevitable: Legal and Policy Issues Related to Space Debris Mitigation and Remediation

Lucy Stewardson

The Brussels Bar, Brussels, Belgium

Steven Freeland

Western Sydney University School of Law, Sydney, Australia

CONTENTS

T HE increasing proliferation of space debris represents a major and pressing challenge, given the threats it poses to space activities as well as to the environment both in space and on Earth. However, as seems to be increasingly the case for activities characterised by rapid technological development, there is no binding legal framework currently in place to comprehensively address the issue. Article IX of the 1967 Outer Space Treaty only briefly touches upon environmental aspects of space, focusing on harmful contamination and the prevention of harmful interference with the space activities of States. More specific and technical standards for the mitigation of space debris have been established in non-binding and voluntary guidelines adopted by the Inter-Agency Space Debris Coordination Committee (IADC) and the United Nations Committee on the Peaceful Uses of Outer Space (COPUOS), complemented by the text of guidelines recently agreed upon by COPUOS aimed at promoting the long-term sustainability of outer space activities. In addition to mitigation standards, COPUOS is also addressing the issue of other legal mechanisms relating to 'space debris and remediation measures', but thus far has not reached a conclusion as to what is possible and practicable. Debris mitigation and, in particular, remediation measures raise many technical, economic and political questions. The legal aspects of remediation also appear challenging: essential matters such as control and jurisdiction over space objects, liability in case of damage during remediation operations, and intellectual property regarding sensitive data and technology, each raise complex questions that require careful consideration as to the most appropriate mechanisms for regulation. This chapter proposes to examine some of the primary legal and policy challenges – and opportunities – related to the regulation of space debris, in terms of both mitigation and remediation.

6.1   INTRODUCTION

Outer space is a critical area upon which the world has grown increasingly dependent. Already during the Cold War, its highly strategic character was acknowledged, with both the United States and the Soviet Union striving to demonstrate their superiority by gaining access to outer space in a fierce competition that came to be known as the 'space race' [1, 2]. The rapid evolution of technology since that period has led outer space to play a key part in most human activities.

Along with the increase in use of, and reliance upon, outer space, the multiplication of space objects has resulted in heavy pollution of the space environment, threatening the future use of space. Space debris has become a major concern within the international community, requiring an urgent and coordinated response, in terms both of mitigation and of remediation of space debris.

This chapter will first emphasise the main elements of the issue of space debris through an overview of the current factual situation and the threats it entails (Section 6.2). Existing treaty law will then be examined in order to understand the current legal framework relating to – and its inadequacy to regulate – the question of space debris (Section 6.3). Non-binding, technical standards have emerged regarding debris mitigation and must continue to develop (Section 6.4). On the other hand, remediation efforts, although indispensable, still give rise to unresolved and complex legal issues. These questions – and possible approaches for solving them – will be examined in the final section (Section 6.5).

6.2   SPACE DEBRIS: A PRESSING ISSUE

Before delving into the intricacies of the current legal framework, as well as its lacunae and opportunities, it is necessary to provide some factual background and context regarding space debris.

In spite of the existence of several instruments aiming to regulate space debris at national, regional and international levels, there exists no universally accepted nor legally binding definition of space debris in international instruments [3]. The United Nations Committee on the Peaceful Uses of Outer Space (COPUOS) is considered as the main forum for issues relating to activities in outer space. In 2007, COPUOS adopted the Space Debris Mitigation Guidelines. These define space debris as:

> All man-made objects, including fragments and elements thereof, in Earth orbit or re-entering the atmosphere, that are non-functional [4].

This definition was drawn from a set of guidelines adopted by the Inter-Agency Space Debris Coordination Committee (IADC) in 2002 [5] and was accepted by the UN General Assembly as part of the endorsement of the COPUOS Space Debris Mitigation Guidelines as a whole [6, Para. 26].

Traditionally, space debris has been broken down into four categories: inactive payloads, operational debris, fragmentation debris, and micro-particulate matter.[1] The first category, inactive payloads, comprises satellites and spacecraft which have become inoperative after a mission and are no longer controlled by operating entities on Earth. Operational debris, on the other hand, refers to mission-related objects and parts of spacecraft which are released during the launching phase and discarded when

---

[1]It must be noted that different categories of space debris potentially give rise to different legal issues, in terms of definition or liability for instance.

they are no longer in use. In-orbit break-ups caused by explosions, collisions and accidents in space, lead to the creation of fragmentation debris. Finally, micro-particulate matter arises from shedding from the surface of in-orbit objects confronted with the extremely hostile environment of outer space [7, pp. 3–8].[2]

Major debris-generating events, such as collisions and intentional destructions of satellites, have further contributed to creating a significant increase in the number of space debris.[3]

In January 2019, the European Space Agency estimated that 34,000 space debris objects larger than ten centimetres were currently gravitating in orbit. The number swells to 900,000 for objects between one and ten centimetres, reaching 128 million regarding objects from one millimetre to one centimetre [14].

The threats posed by space debris are widely acknowledged and this chapter will not delve into the details of the hazards created by the current situation. However, it is worth noting that these threats have been recognised in the 2007 Space Debris Mitigation Guidelines, in which COPUOS stressed that:

> [as] the population of debris continues to grow, the probability of collisions that could lead to potential damage will consequently increase.[4]

This refers to the concern, shared by many States and experts, of a vicious circle in which increasingly frequent incidents would in turn generate more space debris, which would further increase the risk of damage to spacecraft [12, p. 802]. The exponential proliferation of space debris had already been predicted in 1978 by NASA scientists D. J. Kessler and B. G. Cour-Palais. They pointed out the risk of creating a 'debris belt' around Earth as the frequency of collisions increased, eventually rendering outer space nearly un-navigable [15, pp. 2637–2646].

The COPUOS Space Debris Mitigation Guidelines also highlight a second significant threat resulting from the growing population of space debris, namely the 'risk of damage on the ground, if debris survives Earth's atmospheric re-entry'.[5] Although

---

[2]See also Gupta, V. "Critique of the International Law on Protection of the Outer Space Environment." [8, p. 21],Viikari, L. "The Environmental Element in Space Law: Assessing the Present and Charting the Future." [9, pp. 35–36]

[3]These events include the destruction by China of its own satellite (Fengyun-1C) in 2007, followed by a similar operation by the United States of America on one of its own satellites and, in March 2019, the intentional destruction by India of its own spacecraft. The Iridium 33-Kosmos 2251 collision in 2009 also resulted in the creation of thousands of debris. See National Aeronautics and Space Administration (2014, January), "Fengyun-1C Debris Cloud Remains Hazardous." [10, pp. 2–3], National Aeronautics and Space Administration (2019, August), "Two Breakup Events Reported." [11, pp. 1–2]. See also McCormick, P. K. "Space debris: Conjunction opportunities and opportunities for international cooperation", *Science and Public Policy, 40(6)*. [12, pp. 801–802], Freeland, S. "The 2008 Russia/China Proposal for a Treaty to Ban Weapons in Space: A Missed Opportunity or an Opening Gambit?" [13, pp. 261–271].

[4]Chapter 1 of the Space Debris Mitigation Guidelines of the Committee on the Peaceful Uses of Outer Space [4].

[5]Chapter 1 of the Space Debris Mitigation Guidelines of the Committee on the Peaceful Uses of Outer Space [4].

most debris returning to Earth burn up due to the extremely high temperatures, there have already been several accounts of pieces of space debris surviving their journey through Earth's atmosphere and crashing on the ground or in the oceans [9, pp. 40–41]. Such impacts can cause significant damage to property and life on the Earth's surface. The probability of returning debris resulting in personal injury has been estimated to be of the order of one in ten thousand [16].

6.3 THE CURRENT INTERNATIONAL LEGAL FRAMEWORK REGARDING SPACE DEBRIS

Concern about space debris has become an increasingly pressing topic in international fora due to the global consequences of pollution in outer space. However, there is no binding legal framework to satisfactorily and comprehensively address the issue of space debris.

The Treaty on Principles Governing the Activities of States in the Exploration and Use of Outer Space, including the Moon and Other Celestial Bodies (the Outer Space Treaty) [17], deemed to be the backbone of international space law, was drafted in 1967 in the midst of the Cold War and focuses primarily on issues of weaponization, military activities in outer space and cooperation between States [2, pp. 34–37, 44].[6] The environmental aspect of space is only abstractly considered in Article IX of the treaty, which provides that:

> [in] the exploration and use of outer space, including the Moon and other celestial bodies, States Parties to the Treaty shall be guided by the principle of cooperation and mutual assistance and shall conduct all their activities in outer space, including the Moon and other celestial bodies, with due regard to the corresponding interests of all other States Parties to the Treaty. States Parties to the Treaty shall pursue studies of outer space, including the Moon and other celestial bodies, and conduct exploration of them so as to avoid their harmful contamination and also adverse changes in the environment of the Earth resulting from the introduction of extraterrestrial matter and, where necessary, shall adopt appropriate measures for this purpose. (...) [7]

The article then immediately refocuses on the prevention of harmful interference with other States' activities in space through a consultation process.

The first sentence of Article IX establishes the crucial international principle of cooperation and mutual assistance in the context of space activity, which comprises an obligation of 'due regard to the corresponding interests' of other State Parties [8, p. 25]. The only mention of the outer space environment, otherwise largely ignored in the Treaty [18, p. 97], appears in the second paragraph of Article IX, which calls upon States to avoid harmful contamination of outer space (known as forward con-

---

[6]For more background on the adoption of the Outer Space Treaty and development of outer space law, see Jankowitsch, P. [1, pp. 1–28].

[7]Article IX of the Outer Space Treaty [17].

tamination). The article then stresses the necessity to avoid backward contamination, resulting from the introduction of extraterrestrial matter into the Earth's environment. Very little attention is thus devoted to environmental issues in outer space. Moreover, the relevance of Article IX in the context of space debris is uncertain, as there is still discussion amongst scholars as to whether or not space debris must be regarded as falling within the scope of 'harmful contamination'.[8] Some authors contend that the provision of Article IX should be construed as referring only to pollution arising from the release of chemical, biological or radioactive contaminants in outer space, thus not encompassing space debris [8, p. 26] [3, p. 74].

The international conventions adopted subsequently to govern activities of States in outer space do not provide any more help in addressing the issue of space debris. Shortly after the adoption of the Outer Space Treaty in 1967, several instruments were drafted to deal with various aspects of space activities. These are the 1968 Agreement on the Rescue of Astronauts, the Return of Astronauts and the Return of Objects Launched into Outer Space (Rescue Agreement) [20], the 1972 Convention on International Liability for Damage Caused by Space Objects (Liability Convention) [21], the 1975 Convention on Registration of Objects Launched into Outer Space (Registration Convention) [22], and the 1979 Agreement Governing the Activities of States on the Moon and Other Celestial Bodies (Moon Agreement) [23].

Of these four conventions, only the Moon Agreement contains provisions relating to the environment in outer space. Article VII of the Moon Agreement establishes a heightened protection of the space environment by calling upon States to avoid disrupting its existing balance, while Article XI holds natural resources in outer space to be the common heritage of mankind, prescribing their rational management along with the equitable sharing of benefits derived from these resources. This last provision in particular gave rise to reluctance on the part of many spacefaring nations to sign the convention [2,24], rendering the Moon Agreement practically irrelevant with only eighteen ratifications, and none by major spacefaring nations.[9]

Although the Liability Convention does not directly address the issue of space environment and debris as such, it does provide some indications regarding liability for damage caused by space objects[10] and is thus relevant in the event of damage caused by space debris.

According to the Convention, liability falls on the 'launching State', defined as the State which 'launches or procures the launching of a space object' or 'from whose territory or facility a space object is launched'.[11] As a result, there is often more than

---

[8]For a discussion of the meaning of the terms "harmful contamination" in Article IX of the Outer Space Treaty and its scope regarding space debris, see Stubbe, P. [19, pp. 155–167].

[9]The parties to the Moon Agreement, in September 2019, were: Armenia, Australia, Austria, Belgium, Chile, Kazakhstan, Kuwait, Lebanon, Mexico, Morocco, the Netherlands, Pakistan, Peru, the Philippines, Saudi Arabia, Turkey, Uruguay and Venezuela.

[10]Article I(d) of the Liability Convention specifies that, for the purposes of the Convention, the term 'space objects' includes 'component parts of a space object as well as its launch vehicle and parts thereof'.

[11]Article I(c)(i), (ii) of the Liability Convention. It must be noted that in the case of damage caused by space debris, the identification of the launching State(s) can prove difficult as it may

one launching State for the purposes of the Liability Convention [25, p. 471], and Article V of the Convention specifies that all involved launching States shall be jointly and severally liable.[12]

The Convention organises a tiered regime whereby liability is determined depending on the location of the damage. On the one hand, damage caused by a space object on the surface of the Earth or to an aircraft in flight leads to strict, or absolute, liability of the launching State(s).[13] In the case of space debris re-entering the atmosphere and causing damage on Earth, the launching State(s)[14] would incur absolute liability. A launching State may only be exonerated from absolute liability in cases where the damage has resulted from the gross negligence, or an act or omission done with intent to cause damage, on the part of the claimant State or the person(s) it represents, unless the activities from which the damage resulted were in breach of international law. For damage occurring in outer space, on the other hand, liability is fault-based and negligence must be proved.[15] This would for example be the case for in-orbit collisions between two space objects, such as a piece of space debris and an active spacecraft.

Chain reactions are also envisaged in the Liability Convention, which further provides that if damage caused by one State to another State's space object leads to damage to a third party — elsewhere than on the surface of the Earth —, the launching States of both space objects that caused damage to the third party will be jointly and severally liable to the third State.[16]

The Liability Convention establishes a procedural mechanism to deal with compensation claims, first through the diplomatic channel and then, if no agreement has been reached within a year, through the establishment of a Claims Commission.[17] However, the efficiency of this mechanism is compromised by the fact that the decision of the Claims Commission is final and binding only 'if the parties have so agreed' — which, in view of States' reluctance regarding binding dispute resolution mechanisms, is unlikely [25, p. 482]. In the absence of such an agreement, the findings of the Claims Commission merely constitute a recommendatory award which the parties 'shall consider in good faith'.[18]

---

be unclear which particular piece of debris — from a larger cloud of numerous debris — caused the damage, and the source of that piece (see Viikari, L. [24, p. 735], Stubbe, P. [19, pp. 405–406]).

[12] Article V(1) of the Liability Convention. The victim of damage caused by a space object may thus claim full compensation from any of the launching States.

[13] Article II of the Liability Convention.

[14] Article VI(1), (2) of the Liability Convention.

[15] Article II of the Liability Convention (see Kerrest, A., and C. Thro. [26]).

[16] Article IV of the Liability Convention.

[17] Articles IX-VX of the Liability Convention.

[18] Article XIX(2) of the Liability Convention.

6.4 MITIGATION OF SPACE DEBRIS: OVERVIEW OF THE MAIN INTERNATIONAL GUIDELINES

In light of the inadequacy of the international legal framework to effectively address the issue of space debris, and in the face of the growing threat that these debris constitute, a variety of more technical instruments has been adopted by several States and international agencies.[19] At the international level, these instruments most notably include the 2002 Inter-Agency Space Debris Coordination Committee Space Debris Mitigation Guidelines (IADC Guidelines), to a large extent endorsed in 2007 by COPUOS in its own Space Debris Mitigation Guidelines.

In June 2019, COPUOS adopted Guidelines for the Long-Term Sustainability of Outer Space Activities (Long-Term Sustainability Guidelines), which also appear relevant in the context of space debris.

This section will highlight the context and main features of these sets of guidelines (6.4.1) before examining their shortcomings and possibilities for improvement (6.4.2).

6.4.1 Content of the IADC and COPUOS Guidelines

Both the IADC and the COPUOS Space Debris Mitigation Guidelines formulate fundamental standards for mitigation of space debris. Although they are very similar, the two sets of guidelines do differ in some respects. This section will outline the main elements contained in the IADC Guidelines (6.4.1.1) and in the COPUOS Space Debris Mitigation Guidelines (6.4.1.2), before providing an overview of the new Long-Term Sustainability Guidelines (6.4.1.3).

6.4.1.1 The IADC Space Debris Mitigation Guidelines (2002)

Established in 1993, the IADC is a preeminent international governmental forum comprising national and international space agencies aiming to coordinate the agencies' activities with regard to space debris from a technical point of view [28]. The IADC Space Debris Mitigation Guidelines were formally adopted by consensus in October 2002 and were deemed to reflect 'the fundamental mitigation elements of a series of existing practices, standards, codes and handbooks developed by a number of national and international organizations'.[20]

The IADC Space Debris Mitigation Guidelines focus on the following factors relevant for the overall environmental impact of space missions:

(1) limitation of debris released during normal operations;

(2) minimisation of the potential for on-orbit break-ups;

---

[19]For an overview of practices and instruments relating to space debris mitigation, see Legal Subcommittee of COPUOS, "Compendium of space debris mitigation standards adopted by States and international organizations" [27].

[20]Chapter 2 of the Space Debris Mitigation Guidelines of the Committee on the Peaceful Uses of Outer Space [4].

(3) post-mission disposal;

(4) prevention of on-orbit collisions.

Technical recommendations are made in line with those main points of focus. For instance, in order to minimise the risk of on-orbit break-ups, the guidelines recommend the passivation – i.e. the elimination — of all on-board sources of stored energy of a spacecraft or orbital stage when they are no longer required.[21]

Regarding the post-mission disposal of spacecraft when their operational phases are ended, the Guidelines envisage two options depending on the location of the spacecraft. Spacecraft situated in the Geostationary Earth Orbit (GEO)[22] should be re-orbited to another, higher orbit, where they would not interfere with space objects that are still in GEO. The Guidelines provide very technical indications relating to how those manoeuvres could be operated.[23] Spacecraft passing through the Low Earth Orbit Region (LEO)[24], on the other hand, should be de-orbited within 25 years after completion of operations. To this end, the Guidelines specify that direct re-entry into the atmosphere is preferred, but that it should not pose an undue risk to people or property. Manoeuvring spacecraft into an orbit with a reduced lifetime, or retrieving spacecraft at the termination of operations, are also cited as disposal options.[25]

6.4.1.2 The COPUOS Space Debris Mitigation Guidelines (2007)

The Space Debris Mitigation Guidelines adopted by COPUOS in 2007 were largely based on the IADC Guidelines. Although they are non-binding, the COPUOS Space Debris Mitigation Guidelines are considered as the most significant set of international standards relating to space debris; the Guidelines were adopted when the Committee comprised 67 States and, as such, are the expression of a general international consensus among spacefaring nations [30, p. 303]. The significant value and prominent political role of these Guidelines was further enhanced by their endorsement in Resolution 62/217 of the United Nations General Assembly.[26] As a result, a great majority of States refer to, or have in mind as the applicable benchmark standard, the Space Debris Mitigation Guidelines of COPUOS. Examples of the national initiatives

---

[21] IADC Space Debris Mitigation Guidelines, section 5.2.1. (see also McCormick, P. K. [12, p. 806]).

[22] The Geostationary Earth Orbit is an 'Earth orbit (...) whose orbital period is equal to the Earth's sidereal period. The altitude of this unique circular orbit is close to 35,786 km' (IADC Space Debris Mitigation Guidelines, section 3.3.2.).

[23] IADC Space Debris Mitigation Guidelines, sections 5.3.1 and 5.3.2. (see also Viikari, L. [9, pp. 94–95].

[24] The Low Earth Orbit Region (LEO) is a 'spherical region that extends from the Earth's surface up to an altitude of 2,000 km' (IADC Space Debris Mitigation Guidelines, section 3.3.2.)

[25] IADC Space Debris Mitigation Guidelines, section 5.3.2. (see also Hildreth, S. A., and A. Arnold "Threats to U.S. National Security Interests in Space: Orbital Debris Mitigation and Removal." [29, p. 8]).

[26] UN General Assembly. "General Assembly Resolution 62/217, International cooperation in the peaceful uses of outer space." [6] (see Von der Dunk, F. [2, p. 104]).

adopted by States are seen in the latest version of the *Compendium of space debris mitigation standards adopted by States and international organizations.*[27]

The seven Guidelines remain at a generalised level and encourage, on a voluntary basis, actions that would:

(1) limit debris released during normal operations;

(2) minimize the potential for break-ups during operational phases;

(3) limit the probability of accidental collision in orbit;

(4) avoid intentional destruction and other harmful activities;

(5) minimize potential for post-mission break-ups resulting from stored energy;

(6) limit the long-term presence of spacecraft and launch vehicle orbital stages in LEO after the end of their mission;

(7) limit the long-term interference of spacecraft and launch vehicle orbital stages with geosynchronous region after the end of their mission.

These Guidelines thus reflect the main mitigation elements stressed in the 2002 IADC Guidelines. Although the COPUOS Space Debris Mitigation Guidelines are formulated in a broader and less specific manner than the IADC Guidelines, the former make a clear reference to the updated IADC Space Debris Mitigation Guidelines[28] and invite Member States and international organisations to:[29]

> Refer to the latest version of the IADC space debris mitigation guidelines and other supporting documents, which can be found on the IADC website, (...) for more in-depth descriptions and recommendations pertaining to space debris mitigation measures.

6.4.1.3 The Long-Term Sustainability Guidelines (2019)

During its 62nd session in June 2019, COPUOS adopted the Preamble and 21 Guidelines for the Long-Term Sustainability of Outer Space Activities (Long-Term Sustainability Guidelines). These guidelines are grouped into four categories: policy and regulatory framework for space activities; safety of space operations; international cooperation, capacity-building and awareness; and scientific and technical research and development.[30]

Guidelines which are relevant to the issue of space debris include Guideline A.2 of the Long-Term Sustainability Guidelines, which encourages States to 'consider a

---

[27]The latest version of the Debris Mitigation Compendium, dated 25 February 2019, can be accessed at the United Nations Office for Outer Space Affairs (UNOOSA) website [31].

[28]As contained in the annex to UN document A/AC.105/ C.1/L.260.

[29]Chapter 6 of the Space Debris Mitigation Guidelines of the Committee on the Peaceful Uses of Outer Space [4].

[30]COPUOS. Sixty-second session (12–21 June 2019). A/74/20. 2019. Annex II. ('Long-Term Sustainability Guidelines') [32].

number of elements when developing, revising or amending, as necessary, national regulatory frameworks for outer space activities', and more specifically, to:

> [i]mplement space debris mitigation measures, such as the Space Debris Mitigation Guidelines of the Committee on the Peaceful Uses of Outer Space, through applicable mechanisms.

Guideline B.3 directs States to 'promote the collection, sharing and dissemination of space debris monitoring information', while Guideline B.9 calls upon States to 'take measures to address risks associated with the uncontrolled re-entry of space objects', and Guideline D.2 invites them to 'investigate and consider new measures to manage the space debris population in the long term'.

Although the Long-Term Sustainability Guidelines are non-binding, States are called upon to voluntarily implement them through national regulations and mechanisms.[31]

6.4.2   Shortcomings and opportunities regarding space debris mitigation

The COPUOS Space Debris Mitigation Guidelines have been a significant step towards addressing the situation of space debris at the international level. However, they do not cover all aspects of the issue.

One of the main shortcomings of the Guidelines is their silence regarding crucial questions such as liability and insurance [33, p. 37], as well as deliberate destruction of space objects. This last point is particularly contentious, as a complete ban on intentional in-orbit destructions would amount to a restriction of the free use principle embodied in Article I of the Outer Space Treaty [19, p. 438].

Guideline 4 of the COPUOS Space Debris Mitigation Guidelines merely states that [4]:

> The intentional destruction of any on-orbit spacecraft and launch vehicle orbital stages or other harmful activities that generate long-lived debris should be avoided.

However, there is a need for further discussion about the limitation of in-orbit destructions, whether through a complete ban — which seems unlikely — or a more detailed indication of the circumstances under which the deliberate destruction of a space object would be deemed necessary or justified.

Given the fundamental role that the various Guidelines have played, and the increasingly cooperative approach of States regarding the issue of space debris, it is to be hoped that implementation of the Guidelines will continue to improve and that further discussions within COPUOS will allow States to address the key questions mentioned above.

---

[31] Preamble of the Long-Term Sustainability Guidelines.

6.5 REMEDIATION OF SPACE DEBRIS: LEGAL QUESTIONS AND POTENTIAL ANSWERS

In addition to mitigation efforts, there is a growing understanding among States and non-governmental actors that remediation operations — such as active debris removal and on-orbit satellite servicing — are also necessary to properly address the pressing issue of space debris [12, pp. 803–804, 809] [29, p. 10].

However, besides the technical, geopolitical and economic challenges of remediation, remedial efforts also give rise to several legal issues. This section will examine the main legal questions raised by remediation (6.5.1), before highlighting some of the solutions that could contribute to overcoming these difficulties (6.5.2).

6.5.1 Legal hurdles in the way of remediation

The main legal issues that are encountered when considering remediation operations include the definition of space debris (6.5.1.1), jurisdiction and control over the debris (6.5.1.2), liability in case of damage caused during removal operations (6.5.1.3), as well as challenges relating to intellectual property and security (6.5.1.4).

6.5.1.1 Definition of space debris

As discussed above, there is no international legal instrument that defines the term 'space debris'. Although the definition contained in the COPUOS Space Debris Mitigation Guidelines benefits from the widespread acceptance and relative political authority of the Guidelines, it is far from universal or unanimous. This leads to significant difficulties in attempting to determine what constitutes space debris that can be removed from orbit, especially regarding a spacecraft that is no longer operative.

International instruments relating to space debris typically refer to functionality as the relevant criterion to identify space debris, viewing space debris as space objects which have ceased to be functional.[32] However, it has been contended that even seemingly useless and inoperative space objects may constitute valuable assets for a State if, for example, they are 'in reserve' for future operations, or carry classified information [24, p. 719]. It has thus been argued that only the State of registry of a space object, which retains jurisdiction over it even after it ceases to be apparently functional[33], may determine whether that object is non-functional and thus constitutes space debris or not [9, p. 33].[34]

---

[32]This is the case of the definitions of space debris endorsed by IADC, COPUOS and in the European Union Code of Conduct for Outer Space Activities (see Scientific and Technical Subcommittee of COPUOS. "Active Debris Removal – An Essential Mechanism for Ensuring the Safety and Sustainability of Outer Space: A Report of the International Interdisciplinary Congress on Space Debris Remediation and On-Orbit Satellite Servicing." [34, p. 30]).

[33]Article VIII of the Outer Space Treaty states that 'a State Party to the Treaty on whose registry an object launched into outer space is carried shall retain jurisdiction and control over such object'. See Diederiks-Verschoor, I. H. Ph. "Legal Aspects of Environmental Protection in Outer Space Regarding Debris." [35]

[34]See also Scientific and Technical Subcommittee of COPUOS. "Towards Long-term Sus-

6.5.1.2   Jurisdiction and control over space debris

Another consequence of the jurisdiction that the State of registry retains over a space object is the right for that State to exercise sovereign rights over the object [3, pp. 79–80]. As mentioned above, Article VIII of the Outer Space Treaty grants jurisdiction and control of a space object to the State of registry for an indeterminate period of time, even after the object has ceased to be apparently functional.

In this context, 'jurisdiction' is understood as referring to judicial, legislative and administrative competence over a space object [7, p. 68], while 'control' refers to the exclusive legal right to supervise the space objects' activities, in addition to the actual possibility to do so [30, pp. 331–332].

Consequently, in addition to determining whether or not a space object constitutes space debris, the State of registry retains the exclusive right to legally, technically and physically control the object. The view has thus been expressed that removing or otherwise relocating space debris would require the prior consent or authorisation on the part of the State of registry, as doing otherwise would amount to an intrusion into the jurisdiction and control of that State [36, para. 263].[35]

6.5.1.3   Questions of liability

As discussed above, the 1972 Liability Convention organises a regime whereby liability may be incurred jointly and severally by all States which qualify as 'launching States' for the purposes of the Convention. Moreover, damage caused by one State to another State's space object, which then causes damage to a third State, leads the first two States to be jointly and severally liable to the third State. Thus, in the case of remediation:[36]

> If a removal operation causes damage to a third party, the launching States of both space objects (i.e., the removal mechanism and the target object) that caused damage to the third party will be jointly and severally liable under the provisions of the Liability Convention.

Furthermore, the Liability Convention provides for *restitutio in integrum*, which obliges the launching State to provide such reparation as will restore the victim 'to the condition which would have existed if the damage had not occurred'.[37] The State or entity having suffered the damage must thus receive full compensation, for which

---

tainability of Space Activities: Overcoming the Challenges of Space Debris: A Report of the International Interdisciplinary Congress on Space Debris." [33, p. 11]

[35] See also Scientific and Technical Subcommittee of COPUOS. "Active Debris Removal – An Essential Mechanism for Ensuring the Safety and Sustainability of Outer Space: A Report of the International Interdisciplinary Congress on Space Debris Remediation and On-Orbit Satellite Servicing." [34, p. 32], Su, J. [3, pp. 79–80], Hildreth, S. A. and A. Arnold [29, p. 11].

[36] Scientific and Technical Subcommittee of COPUOS. "Active Debris Removal – An Essential Mechanism for Ensuring the Safety and Sustainability of Outer Space: A Report of the International Interdisciplinary Congress on Space Debris Remediation and On-Orbit Satellite Servicing." [34, p. 32].

[37] Article XII of the Liability Convention.

the Liability Convention does not establish a ceiling or maximum amount. Given the technologies and colossal investments involved in space activities, compensation awards for damage caused to space objects can thus become extremely significant and heavy [26, p. 60].

The provisions of the Liability Convention act as significant deterrents for States to carry out remediation operations, as these generally entail perilous manoeuvres such as crossing orbits and re-entering the atmosphere, in the case of active debris removal [3, p. 80] [34, p. 32]. The risk of damage being caused to a space object during remediation operations is thus heightened and, if actually enforced[38], the provisions of the Liability Convention can lead to launching States — even those with little or no control over the operations — being held liable to pay very large amounts of money in compensation for the damage caused.

6.5.1.4   Intellectual property and security aspects of space debris remediation

The fourth major issue that is often highlighted with regards to remediation relates to questions of intellectual property and security.

Remediation operations such as active debris removal and on-orbit satellite servicing imply closely approaching and coming in physical contact with space objects. This requires detailed technical knowledge about the target object, which could be confidential or patented [29, 37].[39] Moreover, once the entity conducting a remediation operation has taken control of the space object, it may potentially access highly sensitive information such as advanced technology or strategic military data. The exchange and treatment of information in the context of remediation is thus a question that needs to be addressed, for instance through the signing of licensing and non-disclosure agreements.

---

[38] As mentioned above, authors have expressed doubts regarding the actual efficacy of the Liability Convention to enforce State liability for damage caused by space objects. This is due, *inter alia*, to the potentially weak value of the decisions by the Claims Commission, as well as the difficulty to prove fault or negligence for damage caused in outer space as there is no systemic space traffic management system to determine which space object is at fault (see Kerrest, A. and C. Thro [26, pp. 66–67], Scientific and Technical Subcommittee of COPUOS. "Active Debris Removal – An Essential Mechanism for Ensuring the Safety and Sustainability of Outer Space: A Report of the International Interdisciplinary Congress on Space Debris Remediation and On-Orbit Satellite Servicing." [34, p. 32].

[39] For further information about concerns on the part of the United States of America relating to the availability of data in cases of debris removal operations, see Scientific and Technical Subcommittee of COPUOS. "Active Debris Removal – An Essential Mechanism for Ensuring the Safety and Sustainability of Outer Space: A Report of the International Interdisciplinary Congress on Space Debris Remediation and On-Orbit Satellite Servicing." [34, pp. 33–35].

6.5.2 Opportunities and possible solutions for remediation

The legal issues raised by remediation activities are complex. However, many propositions have been made to address these and create a legal framework that could support, and even encourage, such operations.

First of all, it has been suggested that the situation requires the adoption of an international convention governing space debris remediation [30, p. 333]. This remediation convention would provide a framework to overcome the key legal issues presented above by defining what constitutes space debris, addressing the question of jurisdiction and control over space objects (through mechanisms facilitating the seeking and granting of permission or the transfer of jurisdiction, for example [34, p. 33]), adapting liability rules to encourage remediation operations, and ensuring that confidential and proprietary data is protected. This can also take the shape of Protocols to the various treaties. For example, it has been argued that it would be desirable to adopt a Protocol to the Liability Convention which would provide for the mitigation of fault if damage occurs in the context of space debris removal operations, or even create an exemption from the scope of application of the Liability Convention [34, p. 32].

Other authors appear even more ambitious and advance the idea of an international treaty pertaining to several aspects of space debris, and not limited to remediation. This would allow the international community to discuss, and reach agreements on, questions relating to the legality of generating space debris, obligations regarding mitigation, collision avoidance, communication of information, and active debris removal [24, p. 758].

Another approach is to build upon the already existing 'soft law' instruments by developing a 'hard law' instrument codifying the emerging principles relating to space debris and environmental issues in outer space. Such principles, agreed upon by States in the continuity of the cooperative approach that has supported the adoption of non-binding instruments, would complement and fill the gaps of the existing corpus of international law [38]. Along the same lines, it has been argued that principles of international environmental law, such as sustainable development, intergenerational equity and due diligence, should be taken into account to provide direction in addressing the challenges of space debris [24, pp. 760-763] [8, p. 38].

In addition, several mechanisms have been envisaged to centralise and exchange information regarding space objects and space debris [39, p. 24], or control space traffic through space traffic management systems [19, pp. 438–439]. The establishment of a 'Global Economic Fund for Space Debris Removal' has also been proposed [34, pp. 28–29][40], which would contribute to the development of remediation technology by providing funding to licensed removal entities. The Fund would be financed by launching States and private actors in proportion to their share of activities in outer space. Such a mechanism could also constitute an appropriate forum for entities active in outer space to coordinate efforts and exchange information regarding remediation.

---

[40]See also Viikari, L. "Environmental aspects of space activities" [24, p. 758].

6.6 CONCLUSION

Despite the widespread acknowledgement of, and concern about, the threats posed by space debris, the legal and policy framework relating to the issue is still 'under construction'.

The conventional framework regarding outer space activities only provides limited help when it comes to space debris. The relevant treaty provisions either appear too vague – such as the unclear threshold, and disputed scope, of 'harmful contamination' of the outer space environment in Article IX of the Outer Space Treaty – or are contained in politically weak instruments – such as the provisions of the Moon Agreement relating to the protection of the outer space environment. The Liability Convention touches upon questions of liability in the case of damage caused by space debris, but none of the five United Nations treaties on outer space satisfactorily address the issue of the creation and multiplication of space debris.

As a result, non-binding instruments have been adopted at the international level to guide States in the right direction by establishing fundamental standards for the mitigation of space debris. In spite of their shortcomings, the IADC Guidelines and the COPUOS Space Debris Mitigation Guidelines contain essential elements of mitigation and benefit from a high level of implementation. Furthermore, the recent adoption of the Long-Term Sustainability Guidelines highlights the ongoing improvement and expansion of instruments dealing with mitigation. However, further discussions and developments are needed, as some key issues — regarding liability and intentional destruction of satellites, for instance — must still be addressed.

In addition to mitigation, remediation efforts must be carried out in order to deal with the pressing threat of space debris. Remediation gives rise to many technical, financial and political challenges, as well as legal difficulties. Some crucial questions that remain to be answered include the definition of what constitutes space debris; what mechanisms would allow the removal of debris without violating States' jurisdiction over these; the liability which may arise in case of damage caused during remediation operations; and sensitive questions relating to intellectual property and national security. However, many ideas and propositions have been presented in order to address these legal issues and enable the conduct of much needed remediation operations.

Although improvements must still be made in terms of mitigation of space debris, and fundamental questions must be answered regarding remediation, the constant developments and efforts that can be observed at the national and international levels allow for a certain optimism. States and, more broadly, all actors involved in one way or another in space activities, are growing increasingly aware of the fact that it really is a matter of addressing the inevitable, and that this can only be achieved through cooperation and effective action.

ACRONYM

IADC Inter-Agency Space Debris Coordination Committee

COPUOS United Nations Committee on the Peaceful Uses of Outer Space

UNOOSA United Nations Office for Outer Space Affairs

NASA National Aeronautics and Space Administration

GEO Geostationary Earth Orbit

LEO Low Earth Orbit Region

GLOSSARY

Outer Space Treaty: Treaty on Principles Governing the Activities of States in the Exploration and Use of Outer Space, including the Moon and Other Celestial Bodies (1967).

Rescue Agreement: Agreement on the Rescue of Astronauts, the Return of Astronauts and the Return of Objects Launched into Outer Space (1968).

Liability Convention: Convention on International Liability for Damage Caused by Space Objects (1972).

Registration Convention: Convention on Registration of Objects Launched into Outer Space (1975).

Moon Agreement: Agreement Governing the Activities of States on the Moon and Other Celestial Bodies (1979).

IADC Guidelines: Inter-Agency Space Debris Coordination Committee Space Debris Mitigation Guidelines (2002).

UNCOPUOS (Mitigation) Guidelines: Space Debris Mitigation Guidelines of the United Nations Committee on the Peaceful Uses of Outer Space (2007).

Long-Term Sustainability Guidelines: Guidelines for the Long-Term Sustainability of Outer Space Activities of the United Nations Committee on the Peaceful Uses of Outer Space (2019).

Geostationary Earth Orbit (GEO): 'Earth orbit (...) whose orbital period is equal to the Earth's sidereal period. The altitude of this unique circular orbit is close to 35,786 km' (IADC Space Debris Mitigation Guidelines, section 3.3.2.).

Low Earth Orbit Region (LEO): 'Spherical region that extends from the Earth's surface up to an altitude of 2,000 km' (IADC Space Debris Mitigation Guidelines, section 3.3.2.).

FURTHER READING

von der Dunk, F. and Tronchetti, F. (2015). *Handbook of Space Law.* Cheltenham: Edward Elgar Publishing.

Jakhu, R.S. and Dempsey, P.S. (2017). *Routledge Handbook of Space Law.* New York: Routledge.

Viikari, L. (2008). *The Environmental Element in Space Law: Assessing the Present and Chartering the Future.* Leiden: Martinus Nijhoff.

Bibliography

[1] P. Jankowitsch. The background and history of space law. In *Handbook of Space Law, by F. von der Dunk and F. Tronchetti*, pages 1–28. Cheltenham: Edward Elgar Publishing, 2015.

[2] F. von der Dunk. International space law. In *Handbook of Space Law, by F. von der Dunk and F. Tronchetti*, pages 29–126. Cheltenham: Edward Elgar Publishing, 2015.

[3] J. Su. Control over activities harmful to the environment. In *Routledge Handbook of Space Law, by R. S. Jakhu and P.S. Dempsey*, pages 73–89. New York: Routledge, 2017.

[4] UNCOPUOS. "Space Debris Mitigation Guidelines of the Committee on the Peaceful Uses of Outer Space." Report of the Committee on the Peaceful Uses of Outer Space on its Fiftieth Session (6-15 June 2007), GAOR, Sixty-second session, Supp. No. 20, A/62/20. 2007. Annex IV.

[5] Inter-Agency Space Debris Coordination Committee. "Space Debris Mitigation Guidelines." United Nations Office for Outer Space Affairs. September 2007. IADC Space Debris Mitigation Guidelines, (accessed on September 4, 2019): http://www.unoosa.org/documents/pdf/spacelaw/sd/IADC-2002-01-IADC-Space_Debris-Guidelines-Revision1.pdf.

[6] UN General Assembly. "General Assembly Resolution 62/217, International cooperation in the peaceful uses of outer space." A/RES/62/217. 21 December 2007.

[7] H. A. Baker. *Space Debris: Legal and Policy Implications*. Vol. 6. Dordrecht: Martinus Nijhoff, 1989.

[8] V. Gupta. Critique of the international law on protection of the outer space environment. *Astropolitics*, 14:20–43, 2016.

[9] L. Viikari. *The Environmental Element in Space Law: Assessing the Present and Charting the Future*. Leiden: Martinus Nijhoff, 2008.

[10] National Aeronautics and Space Administration. Fengyun-1c debris cloud remains hazardous. In *Orbital Debris Quarterly*, pages 2–3, Vol. 18–1, January 2014.

[11] National Aeronautics and Space Administration. Two breakup events reported. In *Orbital Debris Quarterly*, pages 1–2, Vol. 23–3, August 2019.

[12] P. K. McCormick. Space debris: Conjunction opportunities and opportunities for international cooperation. *Science and Public Policy*, 40:801–813, 2013.

[13] S. Freeland. The 2008 Russia/China Proposal for a Treaty to Ban Weapons in Space: A Missed Opportunity or an OpeningGambit? In *Proceedings of the International Institute of Space Law 2008: 51st Colloquium on the Law of Outer Space, Glasgow, Scotland*, pages 261–271. AIAA, 2008.

[14] European Space Agency. "Space Debris by the Numbers". Information correct as of January 2019, (accessed on August 6, 2019): https://www.esa.int/Safety_Security/Space_Debris/Space_debris_by_the_numbers.

[15] D. J. Kessler and D. G. Cour-Palais. Collision frequency of artificial satellites: The creation of a debris belt. *Journal of Geophysical Research*, 83(A6):2637–2646, 1978.

[16] International Academy of Astronautics. "Cosmic Study on Space Traffic Management" (accessed on August 13, 2019): https://iaaweb.org/iaa/Studies/spacetraffic.pdf.

[17] *Treaty on Principles Governing the Activities of States in the Exploration and Use of Outer Space, including the Moon and Other Celestial Bodies, Opened for Signature in Washington, London and Moscow, 27 January 1967.* Vol. 610, United Nations Treaty Series (UNTS), 1967.

[18] H. H. Jr. Almond. A draft convention for protecting the environment of outer space. In *Proceedings of the Twenty-third Colloquium on the Law of Outer Space. Tokyo, Japan*, pages 97–102, 1980.

[19] P. Stubbe. *State Accountability for Space Debris. A Legal Study of Responsibility for Polluting the Space Environment and Liability for Damage Caused by Space Debris (Studies in Space Law).* Leiden: Martinus Nijhoff, 2018.

[20] *Agreement on the Rescue of Astronauts, the Return of Astronauts and the Return of Objects Launched into Outer Space, Opened for Signature in London, Moscow and Washington, 22 April 1968.* Vol. 672, United Nations Treaty Series (UNTS), 1968.

[21] *Convention on International Liability for Damage Caused by Space Objects, Opened for Signature in London, Moscow and Washington, 29 March 1972.* Vol. 962, United Nations Treaty Series (UNTS), 1972.

[22] *Convention on Registration of Objects Launched into Outer Space, Opened for Signature in New York, 14 January 1975.* Vol. 1023, United Nations Treaty Series (UNTS), 1975.

[23] *Agreement Governing the Activities of States on the Moon and Other Celestial Bodies, Opened for Signature in New York, 18 December 1979.* Vol. 1363, United Nations Treaty Series (UNTS), 1979.

[24] L. Viikari. Environmental Aspects of Space Activities. In *Handbook of Space Law, by F. von der Dunk and F. Tronchetti*, pages 717–761. Cheltenham: Edward Elgar Publishing, 2015.

[25] S. Freeland. There's a Satellite in My Backyard-Mir and the Convention on International Liability for Damage Caused by Space Objects. *UNSW Law Journal*, 24(4):462–484, 2001.

[26] A. Kerrest and C. Thro. Liability for damage caused by space activities. In *Routledge Handbook of Space Law, by R. S. Jakhu and P. S. Dempsey*, pages 59–72. New York: Routledge, 2017.

[27] Legal Subcommittee of UNCOPUOS. "Compendium of Space Debris Mitigation Standards Adopted by States and International Organizations." Fifty-fifth Session (4–15 April 2016), A/AC.105/C.2/2016/CRP.16. 2016.

[28] A. Tuozzi. "The Inter-Agency Space Debris Coordination Committee (IADC): An Overview of IADC's Annual Activities". United Nations Office for Outer Space Affairs. November 2018, (accessed September 4, 2019): http://www.unoosa.org/documents/pdf/icg/2018/icg13/wgs/wgs_23.pdf.

[29] S. A. Hildreth and A. Arnold. *Threats to U.S. National Security Interests in Space: Orbital Debris Mitigation and Removal. Report Prepared for Members and Committees of Congress, Congressional Research Service, 8 January 2014.* 2014.

[30] L. Li. Space Debris Mitigation as an International Law Obligation: A Critical Analysis with Reference to States Practice and Treaty Obligations. *International Community Law Review*, 17(3):297–335, 2015.

[31] UNOOSA. Space Debris Mitigation Compendium, dated 25 February 2019, can be accessed at: http://www.unoosa.org/documents/pdf/spacelaw/sd/Space_Debris_Compendium_COPUOS_25_Feb_2019p.pdf.

[32] UNCOPUOS. "Guidelines for the Long-term Sustainability of Outer Space Activities." Sixty-second Session (12–21 June 2019), UN Doc. A/74/20. 2019. Annex II.

[33] Scientific and Technical Subcommittee of UNCOPUOS. "Towards Long-Term Sustainability of Space Activities: Overcoming the Challenges of Space Debris: A Report of the International Interdisciplinary Congress on Space Debris." Forty-eighth Session (7–18 February), UN Doc A/AC.105/C.1/2011/CRP.14. 2011.

[34] Scientific and Technical Subcommittee of UNCOPUOS. "Active Debris Removal – An Essential Mechanism for Ensuring the Safety and Sustainability of Outer Space: A Report of the International Interdisciplinary Congress on Space Debris Remediation and On-Orbit Satellite Servicing." Forty-ninth Session (6–17 February 2012), UN Doc. A/AC.105/C.1/2012/CRP.16. 2012.

[35] I. H. Ph. Diederiks-Verschoor. Legal Aspects of Environmental Protection in Outer Space Regarding Debris. In *Proceedings of the 30th Colloquium on the Law of Outer Space of the IISL*. Brighton: AIAA, 1987.

[36] UN General Assembly. "Report of the Committee on the Peaceful Uses of Outer Space on its Sixty-first Session (20–29 June 2018)." Seventy-third Session, Supp. No. 20, UN Doc. A/73/20. GAOR, 2018.

[37] M. J. Listner. Legal Issues Surrounding Space Debris Remediation. *The Space Review*, 6(08), 2012.

[38] U. M. Bohlmann and S. Freeland. The Regulation of Space Activities and the Space Environment. In *Routledge Handbook of International Environmental Law, by S. Alam, Md J. H. Bhuiyan and T. M. R. Chowdhury*, pages 375–391. London: Taylor & Francis Ltd, 2012.

[39] UNCOPUOS. "Report of the Legal Subcommittee on its Fifty-seventh Session (9–20 April 2018)." Sixty-first Session (20–29 June 2018), UN Doc. A/AC.105/1177. 2018.

CHAPTER 7

Risk Assessment of Space Activities in Light of Space Debris Issue

Cécile Gaubert

Gaubert Law Firm, Paris, France

CONTENTS

T HE number of space debris is increasing yearly as space activities, launching and operating space objects, are also growing especially with the development of the "new space" that includes small satellites deployment or new activities such as services in orbit. There are various reasons for space debris creation: launch operations, loss of control of a space object or even intentional destruction of a satellite. With respect to risks associated with space debris, even though there is, as of today, no significant loss history related to damages caused by space debris, some incidents have occurred either on Earth or in orbit. Furthermore, there are ongoing projects aiming to clean space. Legal and risk issues are also associated with these projects. This leads to the fact that space debris is an issue in terms of legal risks and risk assessment. As of today, there is no legal binding regulation dedicated to space debris at the international level, even though we might find some national regulation. Therefore, this chapter will discuss the legal issues and risk assessment along with potential transfer to a dedicated market (such as insurance) or States. This chapter will not only focus on risks created by space debris, but also risks associated with remediation projects, that can create new risks. Finally, this chapter will review the paths that could be followed by the insurance market to support the development of space debris remediation projects.

7.1 INTRODUCTION

Since 1957, the date of launch of the first manmade satellite Sputnik 1, more than 5,000 launches carrying one or more satellites have been performed. The result of these launches is that, as of the beginning of year 2019, roughly 9,000 satellites have been put into orbit. There are 5,000 satellites orbiting with only 2,062 in operation. There is a high number of non-operational satellites still orbiting, being qualified as space debris. But space debris are not only non-operating satellites and can be of various natures.

Space debris are generated by different and multiple sources that can be classified as follows:

- Non operational satellites: the satellites that have reached their contractual end of life or that have sustained a prematured failure leading to a loss of control.

- The launchers upper stages remaining in orbit once the launched is performed (being successful or not). The lower stages being designed to re-enter the atmosphere are also at the origin of debris.

- Debris associated with missions: these are objects or materials detaching themselves from the space object, as for example, pyrotechnic items or the payload's adapters etc.

- Debris created by the payload or launchers' explosions.

- Collision in orbit between space objects that creates debris, such as the collision between Iridium 33 and Kosmos 2251 in 2009.

- Voluntary destruction of satellites. In this respect, we may note that the United States in 2008, China in 2007 and very recently India in 2019 have destroyed one of their satellites using a ballistic missile fired from Earth. The Russian Federation does possess the equivalent technology without having used it on satellites. This destruction of satellites positioned in low earth orbit creates a wide number of debris in outer space and therefore increases the risks of collision between an active space object and a debris. Indeed, as a matter of example, in January 2013 a small Russian satellite, Blits, was hit by a piece of space debris, generated by the destruction of the Fengyun-1C Chinese satellite back in 2007. It is to be noted that there have been no legal consequences further to this collision.

The space debris risk is a growing risk inherent to space activities, so there is a real need to find ways to manage said risks and to provide solutions to both mitigate risks and manage them.

This chapter will discuss the legal framework relating to space debris and more specifically the issues linked to legal and technical risks (7.2). Based on the legal environment, it will be necessary to discuss the involvement of the space insurance sector to assess possible ways to transfer space debris risks to the space insurance market (7.3) and to explore paths under which the insurance market may support debris removal projects (7.4).

7.2   RISKS LINKED TO DEBRIS

When asserting space debris risks, the very first issue relates to the fact that there is no legal definition of space debris. Nevertheless, the United Nations Treaties (Outer Space Treaty [1] and Liability Convention [2]) are clear on the fact that States are internationally liable for space objects being defined as the object itself and its components parts[1]. Due to the lack of legal definition, there have been some initiatives to define space debris. As such, we may refer to the UNCOPUOS Space Debris Mitigation Guidelines, where space debris are defined as *all manmade objects, including fragments and elements thereof, in Earth orbit or re-entering the atmosphere, that are non functional* [3, p. 1, pt. 1]. It is worthwhile to highlight at this stage, that this definition is not legally binding. Even though there is no legal definition of space debris, the issue of liability in the case of damages caused by space debris shall be nevertheless discussed in order to assess the risks attached to such debris and the eventual risk management associated to them.

Therefore, we shall ask ourselves to what extent a liability is attached to space debris and what are the consequences on the liability of a State or space operator (pri-

---

[1] Article VII of the OST: Each State Party to the Treaty that launches or procures the launching of an object into outer space, including the Moon and other celestial bodies, and each State Party from whose territory or facility an object is launched, is internationally liable for damage to another State Party to the Treaty or to its natural or juridical persons by such object or its component parts on the Earth, in air space or in outer space, including the Moon and other celestial bodies.

vate or not) when damages are caused due to space debris, whatever their definition is.

In this section, we will assess in the first part the legal context surrounding space debris risks and whether there is any liability linked to damages caused by space debris (7.2.1). In the second part, we will focus on major risks inherent to space debris (7.2.2) and in the third part, we will discuss the specificities in respect to space debris removal projects (7.2.3).

7.2.1   Legal environment of space debris

There are various legal grounds at international or national levels on which we can rely to assess the legal framework for space debris.

7.2.1.1   International regulations

The international space regime provides for the concepts to be used in order to assess the liability of a State.

The issue of liability at the international level is managed by both the Outer Space Treaty and the Liability Convention. Important articles are II, III, and IV of the Liability Convention, which impose absolute liability[2] or fault-based liability[3], depending on the place of occurrence of the damage. Article IV on its side provides for the conditions of joint liability between launching states[4].

---

[2]Art. II of the Liability Convention: A launching State shall be absolutely liable to pay compensation for damage caused by its space object on the surface of the earth or to aircraft flight.

[3]Art. III of the Liability Convention: In the event of damage being caused elsewhere than on the surface of the earth to a space object of one launching State or to persons or property on board such a space object by a space object of another launching State, the latter shall be liable only if the damage is due to its fault or the fault of persons for whom it is responsible.

[4]Art. IV of the Liability Convention:

1. In the event of damage being caused elsewhere than on the surface of the earth to a space object of one launching State or to persons or property on board such a space object by a space object of another launching State, and of damage thereby being caused to a third State or to its natural or juridical persons, the first two States shall be jointly and severally liable to the third State, to the extent indicated by the following:

    (a) If the damage has been caused to the third State on the surface of the earth or to aircraft in flight, their liability to the third State shall be absolute;

    (b) If the damage has been caused to a space object of the third State or to persons or property on board that space object elsewhere than on the surface of the earth, their liability to the third State shall be based on the fault of either of the first two States or on the fault of persons for whom either is responsible.

2. In all cases of joint and several liability referred to in paragraph 1 of this article, the burden of compensation for the damage shall be apportioned between the first two States in accordance with the extent to which they were at fault; if the extent of the fault of each of these States cannot be established, the burden of compensation shall be apportioned

Based on the above referenced articles, it can be assessed that in case of damage occurring on earth or in the airspace, the liability of a launching State is qualified as absolute liability, meaning that the victim of the damage doesn't have to prove any fault to make a claim against the Launching State. However, for any damage occurring elsewhere than on earth and in the airspace, the liability of a launching State can only be established if it can be demonstrated that the State has committed a fault and that this fault is the cause of the damage. So both a fault and a causation link must be proven, along with the ability to identify the liable party. The following developments apply solely to damages occurring in-orbit.

7.2.1.1.1 *The fault* The notion of fault is not defined in the Liability Convention, so there is a need to rely on other concepts in order to understand this notion in light of space debris risks.

Article III of the Outer Space Treaty provides for the possibility to make application of international law to promote international cooperation[5]. Therefore, we may use the concepts highlighted under international law to space activities. Henceforth, international law uses the concept of "internationally wrongful act". The violation of an international obligation is then manifested by a failure qualified by action or abstention behaviour; this behaviour can therefore be qualified by a deficit of the State. Article XIII of the draft international responsibility of States by the International Law Commission provides that the act of the State does not constitute a breach of an international obligation unless the State is bound by that obligation at the time when the act happens. It can therefore be a customary or conventional norm since there is no hierarchy of norms in public international law. The main issue, however, will relate to the interpretation of this concept. It is therefore appropriate to wonder about certain points when the question arises as to whether a State has breached an obligation imposed by a primary rule[6].

If we apply this concept to space law and more particularly space debris, we have first to determine if there is any international obligation linked to space debris. At this stage, without any dispute resolution on this issue, we have to rely on the various tools at our disposal, in particular the guidelines and other tools of international organizations such as COPUOS or IADC[7]. These tools having acquired a certain value[8]

---

equally between them. Such apportionment shall be without prejudice to the right of the third State to seek the entire compensation due under this Convention from any or all of the launching States which are jointly and severally liable [Emphasis added].

[5]Outer Space Treaty [4], 1967, Article III, "States Parties to the Treaty shall carry on activities in the exploration and use of outer space, including the Moon and other celestial bodies, in accordance with international law, including the Charter of the United Nations, in the interest of maintaining international peace and security and promoting international cooperation and understanding".

[6]International Law Commission, 2001, Draft articles on Responsibility of States for International Wrongful Acts [5].

[7]Article III (Fault Liability) LIAB of Cologne Commentary on Space Law vol. 2, [6].

[8]Arrêté du 31 mars 2011 relatif à la réglementation technique en application du décret du

could serve as a basis for qualifying the violation of an obligation, and therefore a wrongful act.

The COSPAR Planetary Protection Policy can also be used as an example. This Policy categorizes different space missions. Thus, depending on the category in which the mission is located, it is recommended that the operator follows a certain number of protection measures, such as the development of a trajectory plan, post-launch reports, inventories of organic matter, etc. When damage occurs, it would be wise to check whether the operator has followed the measures contained in the COSPAR document.

For that, it is necessary to start from the postulate that an obligation weighs on the State, who should have, on the basis of the tools at its disposal, supervised the operations of its private actors and which would have failed in this obligation. This would then constitute an illegal act.

Another concept we may discuss and rely on to assess any state liability is the one of custom. The flagship document in this area remains the "Space Debris Mitigation Guidelines of the COPUOS" [3]. However, it falls into the category of soft law and does not enjoy binding force. The question that arises here is therefore whether this soft law would have the consequences of creating a customary law as regards the avoidance of space debris, taking into account the fact that most national laws emphasize the environmental dimension of limiting space debris, as per authorization and licensing mechanisms implemented by most countries through their national laws.

There are several criteria to be met to assess that a practice has become a "custom" with all the consequences attached to such a custom.

- The practice must be state, constant and general. The general tendency tends well to the limitation of space debris. States rely on the IADC guidelines to do this.

- Opinio Juris. This aspect is sometimes difficult to qualify, but in the case of space debris, the desire to be bound by the principle of limiting space debris might be easy to demonstrate since national laws directly or indirectly refer to the guidelines of the IADC.

- Psychological element related to the practice. This element is also verified by the practice of the States which tends more and more to limit the generation of space debris. Some States even go further with the creation of programs to clean up outer space.

The above statement shall be confirmed or not in the future in the context of an eventual dispute resolution.

There is an other concept in international law that could be used in to assess a fault which is referred to as due diligence. This due diligence is characterized when a State takes the necessary measures to avoid the occurrence of a damage. Thus, we could determine the fault of a State which did not take the necessary measures to avoid the

9 juin 2009 relatif aux autorisations délivrées en application de la loi du 3 juin 2008 relative aux opérations spatiales, articles 26 et 43.

creation of space debris. We are then facing a kind of obligation of means to take the necessary measures for the preservation of the space environment on the customary legal basis which could give rise to a general consensus if most of the States were to do so.

It should be recalled that Article IX of the Space Treaty establishes the principle of "due regard" [6, p. 175]. States must undertake their activities by pursuing a duty of cooperation and assistance, taking into account the interests of other States. The tools mentioned above only implement this article. It must be emphasized in this respect that this concept hasn't been raised by the States further to the intentional destruction of satellites from earth performed by the United States, China and India.

The second imperative notion to assess liability under international regulation, in case of damages occurring in-orbit, is the notion of causation between the fault and the damage.

7.2.1.1.2 The causation link As indicated above, a space operator can only be held liable if it has committed a fault and that this fault is at the origin of the damage to a third party. This is called the causation link. This link is not easy to prove when considering the space environment. Indeed, in some cases, there can be a chain of reaction, in which case, several States or operators might be involved, therefore, assessing the link between the fault original fault and the damage becomes difficult.

7.2.1.1.3 Who is liable? The issue here is to identify the entity that can be held liable in case of damages caused by space debris. This is a critical issue as it is sometimes difficult or impossible to identify the liable entity. The Liability Convention holds liable the "launching State" and provides for a definition of a Launching State[9].

When applying this to space debris, it implies that the debris shall not only be monitored, but that the launching State shall also be identified. Due to the size of some debris, it can be in certain situations impossible to identify the launching State.

Concerning damages occurring in orbit, although there are some legal grounds at the international level to rely on to demonstrate a fault, the absence of any case law on this matter makes it difficult for us today to assess exactly the extent of the exposure and liability that can be borne by States in respect of space debris risks.

7.2.1.2 The application of domestic space legislation

Most of national space legislations provides for an authorization or licensing procedure. Under such a procedure, the operator willing to have an authorization shall follow requirements set up by the State. For most of space legislation, space debris are not assessed as such, but the operators shall however meet mitigation criteria.

---

[9]Liability Convention [2] Art. I c): The term "launching State" means:

  i. A State which launches or procures the launching of a space object;

  ii. A State from whose territory or facility a space object is launched.

Some States have developed debris mitigation standards to be followed by the space operators[10].

In the absence of a dedicated and binding international regulation of space debris, national and international space agencies have established some codes of conduct to be applicable to private entities but not mandatory. Here are some examples :

- "Safety Standard NSS-1740.14 – Guidelines and Assessment Procedures for Limiting Orbital Debris", 1995 – NASA;

- "Space Debris Mitigation Standard NASDA-STD-18", 1996 – NASDA;

- "CNES Standards Collection, Method and Procedure Space Debris – Safety Requirements (RNC-CNES-Q40-512)", 1999 – CNES;

- "European Code of Conduct", 2004 – ESA.

However, as an attempt to have a more biding legislation, there is the French Law on Space Operation[11] which refers directly in its Arreté on technical regulation[12] to the COSPAR Planetary Protection Policy[13].

7.2.2   Risks linked to space debris

Whatever their size, space debris may generate damages even though they are relatively small, due to the speed at which the debris is moving into outer space. Space debris are monitored by space agencies and we may say that the number of debris objects estimated by statistical models to be in orbit is: 34,000 objects >10 cm; 900,000 objects from 1 cm to 10 cm and 128 million objects from 1 mm to 1 cm [8].

There are risks linked to space debris evolving in outer space and creating a risk of collision with an active space object (satellites or even worse to the International Space Station), but there are also risks associated with the re-entry of debris into the atmosphere back to earth. We have a very limited history of space debris accidents on earth. There is of course the well-known re-entry of Kosmos 954 that crashed on the Canadian Arctic in 1979 and that thankfully did not harmed the population. However, this crash led to a nuclear leak and negotiations took place between the USSR and Canada to adjust the damages and the associated indemnification. With respect to bodily injury, as of today, only one human being has been hit in 1997 (22 January) by a fifteen centimetres space debris while walking in a park. In this case, no legal case has been brought neither in front of national jurisdiction, nor international arbitration court.

---

[10]See Hakeem Ijaiya - Space Debris: Legal and Policy Implications. Environmental Pollution and Protection, Vol. 2, No. 1, March 2017 [7].

[11]Loi n°2008-518 du 3 juin 2009 relative aux opérations spatiales, article 4, décret n°2009-643 du 9 juin 2009 relatif aux autorisations délivrées en application de la loi du 3 juin 2008 relative aux opérations spatiales, chapitre 1er.

[12]Arrêté du 31 mars 2011 relatif à la réglementation technique en application du décret n°2009-643 du 9 juin 2009 relatif aux autorisations délivrées en application de la loi n°2008-518 du 3 juin 2008 relative aux opérations spatiales.

[13]Articles XXVI and XLIII.

Today, the risk of collision of a spacecraft with an object greater than 1 cm (which could cause the complete destruction of the spacecraft and generate other debris) is estimated at an event every 3 or 5 years. In addition, there were approximately 750,000 dangerous objects in space. With the proliferation of satellite constellation projects and the emergence of new launchers, the launch of more than 1,000 satellites is planned in the next decade; if nothing is done to change current practices, the number of debris is expected to increase in even greater proportions, which would increase the risk of collision. In this context, the actors of the space sector would have to reassess their exposure to the potential damage resulting from collisions and especially the community of insurers. This is why it seems essential to consider measures to contain this risk in an international legal framework.

We have briefly detailed above the risks that are linked to damages that space debris can cause. However, due to the proliferation of projects associated with debris removal, there are risks linked to said projects and that need to be analysed as well.

7.2.3 Risks in respect to debris removal projects

There are several projects relating to space debris removal. Such projects also carry some risks inherent to this mission. These projects may include the use of a servicing satellite, for instance, to de-orbit or remove a non-functioning satellite or even to clean outer space from orbiting debris. What kinds of damages can be caused?

During the course of a removal mission, the satellite performing debris removal can be damaged due to various causes. It can be due to internal failure or collision with another satellite or even with the satellite to be removed. In this case, the operator of the satellite may bear some prejudice due to the damages caused to its satellite.

The satellite that is the subject of the removal mission can also be damaged due to the mission, for example, in the case when a docking phase isn't performed nominally. In this situation, the issue is to know whether the operator/owner of the removed satellite sustains any prejudice following damages to its satellite. We may have some doubts concerning a non-functioning satellite where it might be highly difficult to assess any indemnifiable loss for the owner/operator.

There might also be some risks of damages to a third party, either in orbit or on earth. Indeed, the removal mission may not be going nominally and may be followed by a collision with an in-orbit active satellite. In this case, the liability of either the operator of the satellite performing the removal or the owner/operator of the removed satellite may be called upon, but this is only the case if a fault can be proved against one or both of the operators. The mission could also consist of de-orbiting a non-functioning satellite and having it removed to be burnt while returning in the atmosphere. In this case, both satellites, the one performing the re-entry and the removed satellite may cause damages to third parties on earth or in the airspace. In this context and based on the Liability Convention, no fault shall be proven by the third party victim. Furthermore, based on the Liability Convention, the victim may claim against both launching states.

Some damages due to interference during the performance of the mission can also be considered. In this case, it will be necessary to revert to applicable interference regulation to be able to make a claim.

Based on the above development, we have seen that there are some risks inherent to space debris and the issue is now to identify any possibility to transfer the risks to the space insurance market.

7.3   TRANSFER OF RISKS ON INSURANCE MARKET

With the increasing privatization of space activities, it is crucial to be able to accurately determine liability issues in respect to these activities, but also to financially secure space projects. As a result, insurance has become a major topic in the conduct of these activities.

While the insurance market has developed a long and rich experience in other sectors, such as land, sea or air transport, the specificities of space activities have involved implementation of substantial adaptations to traditional insurance and even introduced new insurance practices.

Before assessing the application of space insurance to space debris (7.3.2), it is necessary to understand the existing space insurance coverage (7.3.1).

7.3.1   Available space insurance coverage

To be able to perform a space project, a certain number of actors is needed. These actors bear risks that are unique to them. Thus, for the various phases of risks including manufacturing, storage, transportation, launch, satellite operations, there are responsibilities identified and specific to each actor. These responsibilities are associated with insurance solutions, which in some cases have been specifically set up for space related risks.

The development of space insurance has coincided with the privatization and commercial development of satellite launch and operation activities. This development concerns not only damage insurance for satellites or launcher, on ground or in outer space, but also liability insurance for space operators, manufacturers, equipment manufacturers, suppliers, etc. In general, we can say that there are two main categories of space insurance, first party property damage insurance for space assets and space liability insurance for damages caused to a third party.

7.3.1.1 *Property damage insurance*

Traditionally, in the context of first party property damage insurance for space assets, three phases of risks are to be counted: on ground, during launch and during life in orbit. For these phases, policyholders, risks, guarantees and insurers will not be the same.

We will focus on launch and in-orbit phases and not on the ground phase, where risks linked to space debris aren't relevant.

"Launch" insurance policies are in force from launch of a space object i.e. when the launch is deemed irreversible. During this launch phase, only satellites are covered, launchers and other shuttles are not directly insured.

During the orbiting phase of space objects, satellites, and mainly commercial satellites, can be covered, from the end of launch up to the end of their contractual life.

The duration of such first party property damage insurances varies from a few days to one year or several years, even to the whole lifetime of a satellite. For launch and operation phases in orbit, satellites are insured for any total, partial or constructive total loss. These first party property insurances are designed to cover, according to the loss formulas provided for in these policies, any loss of control, destruction, impossibility of reaching the specified orbit, but also the cases of reduction of the operational capacity or the life of the satellite, occurring during the warranty period[14].

In principle, launch and in orbit policies provide coverage on an all risks basis, which is why they are called "all risk policies except". Thus, only the exclusions specifically indicated in the policy may be invoked by insurers in order to defeat the guarantee. It will therefore be up to insurers to prove that an exclusion applies and not to the insured to prove that its loss is covered.

These first party property damage insurances for space assets are nowadays well mastered, but they need to adapt themselves to new technologies and new projects under development, such as, for example, satellite constellations, new launchers or debris removal.

Besides the first party property damage there are space liability insurances.

---

[14] The loss cases are as the following:

- TOTAL LOSS being the case where the satellite is totally lost or destroyed or cannot reach its intended orbital position within a certain period of time.

- CONSTRUCTIVE TOTAL LOSS being the case where the operational capacity of a satellite compared to its nominal capacity is above a certain percentage of loss, but the satellite cannot be declared a total loss.

- PARTIAL LOSS being the case where the satellite is partially lost, but cannot be declared a constructive total loss.

7.3.1.2 Third party liability insurances

Satellite launch and in-orbit operations include a high degree of risk of liability for damages caused to a third party resulting from the intended space activity. As such, the risks associated with the launch and the possibilities of damage caused on earth due to the re-entry of the launcher stages were the first concerns of the international community. These concerns had led to the drafting and ratification of the Outer Space Treaty and the Liability Convention dealing in particular with the liability of launching States for damage caused by space objects. In addition to these texts, certain States, which can be qualified as launching states under the Liability Convention and therefore bearing international liability, have decided to legislate on this subject and certain national laws require space operators to insure themselves for the risks involving their liability, as well as the one of the launching State.

The liability insurance policies must therefore respond to possible liability claims, not only under the liability regime provided for in the international treaties and particularly the Liability Convention, but also under national legislations.

Schematically, there are two main categories of third party liability for participants in a space operation: civil liability related to the operation of spacecraft, and third party liability for space products. The latter is underwritten by manufacturers, equipment manufacturers, suppliers, in case of damage caused to a third party due to a defect of the product after delivery.

Third party liability insurance for space objects covers the financial consequences of the liability of an insured for damage caused to a third party as a result of a space insured activity. These insurances are available, in the current state of the market, up to 750 million USD. These insurances are available for launch and in-orbit phases. Traditionally, these insurances are subscribed by the space operator (launching agency or the satellite operator) and the launching states are named additional insured, which means that they will benefit from the coverage (in the amount, conditions and exclusions of the guarantee) in the event a third party decides to make a claim against the launching state. All participants in a launch operation are also generally covered under such insurances (including the manufacturers of the launcher and the satellite(s), and all of their subcontractors and suppliers at whatever level).

The limits of coverage vary according to the legal provisions if they exist or according to the risk apprehension of the operators. The premium associated with this risk is determined by the insurers after an exposure analysis based on various elements related to the activity to be insured, such as launch site used, launch trajectory, details of the impact zones, backup and security procedures, operator and manufacturer's loss history, experience of the launching agency, technical details of the satellites, orbital positioning, planned movements, etc. These insurances have prices that have changed little, with relatively low premium rates for this type of risk. It should be kept in mind that a single major disaster can have a huge impact or even drastically affect the terms and conditions of these coverages. This must be considered by the States in the event of the disappearance of these insurances. Therefore, and as much as possible, policyholders must protect themselves by including contractual provisions that allow them to reduce or waive their liability, through waiver of recourse clauses and other hold harmless protections.

7.3.2 Application to space debris

As there are more and more debris, of various size, orbiting in space, the issue of damages that could be caused by them is a growing concern. The question that we will try to answer now is: are there any insurance schemes applicable to space debris and to what extent would such insurance cover damages caused by said debris?

7.3.2.1 Property damage insurance

Two issues will be discussed in this section, the first one relating to the possible indemnification by insurance in case of property damage caused to a functioning satellite by a debris and the second the indemnification of any damage caused to a satellite performing a removal mission or to a satellite being the subject of a removal mission.

7.3.2.1.1 Damages caused to a functioning satellite by a debris Usually private/commercial satellites operators insure their satellites under launch and in-orbit insurance. As indicated above (7.3.1.1), these insurances will cover on an "all risk" basis the insured satellite in case of damage or failure occurring during the period of validity of the insurance contract. The question that arises is whether such insurance would cover any damages to an insured satellite due to a space debris. Traditionally there is no exclusion in this type of insurance for damages caused by space debris. Therefore, these first party property insurances are designed to cover such damages and will indemnify the operator of the satellite if it is damaged by a debris.

As an illustration that the first party property insurance is covering damages caused by space debris, we may refer to the damage to Ecuador's CubeSat by a space debris. It was launched in April 2013 and one month later in May it collided with a launch vehicle tank debris [9]. As this satellite was insured under first party property damage insurance for the launch phase, it allowed Ecuador's space agency (EXA) to be indemnified for the loss of the satellite[15].

The space insurance market is therefore offering coverage for damages to the insured satellite by space debris, unless a specific exclusion applies. This being said, we have to point out that insurers are however considering the potential exposure of damages caused by space debris when assessing the risk and calculating the premium rate accordingly. This situation may evolve in the future and insurers may analyse more closely risks associated with space debris, due namely to the negative results of the space insurance market for the year 2019.

---

[15]EXA has presented the appropriate claim to the insurance company, which has accepted it and gone forward with the corresponding payment, the space agency said. The insurance payment enabled Ecuador to recoup the nearly USD 800,000 invested in building and launching the Pegaso, according to EXA [10].

7.3.2.1.2 Damages during the performance of a removal mission As indicated above (7.2.2), a satellite performing a debris removal mission may sustain damages during the course of its mission. Based on the existing first party property insurance, it is possible to cover the satellite under said insurance. However said insurance will certainly need to be adapted with respect to terms, conditions, premium rate due to the technical specifications of the satellite and due to its specific mission.

As for damages occurring on the satellite to be removed, in order to identify a possible insurance, it shall first be assessed whether the owner or operator of the removed satellite is sustaining a prejudice while its satellite is being removed. If we are referring to a non-functioning satellite it is very unlikely that the operator can prove any prejudice. If we are referring to a remaining functioning satellite, then the operator, subject to proof of the remaining value of the satellite, may insure its satellite under a first party property insurance. In this specific case, the terms, conditions, premium rate will be the result of negotiation between the owner or operator and the insurers; the latter may wish to have specific conditions and premium rates due to the very specific nature of the mission.

Having discussed the application of first party property insurance to space debris risks, we have to discuss now the application of the existing space third party liability insurances.

7.3.2.2 Third party liability insurance

We will first review the application of space third party liability insurances in the case of damages caused by space debris and secondly in the context of debris removal projects.

7.3.2.2.1 Third party liability insurance and damages caused by a space debris A third-party liability insurance may only be triggered if the liability of the insured can be called upon. This is the reason why the application of this type of insurance is linked to the ability of proving the liability of the party at the origin of a loss. This becomes an issue when the damage has occurred in orbit, where a fault shall be proven as per the Liability Convention.

Another important issue is that the insurance shall be in force when damage to the third party occurs. This is a critical issue when associated with space debris risks. On a standard basis, this insurance is subscribed for a maximum of a twelve month period and shall be renewed each year. In case a functioning space object insured under a third party liability insurance becomes space debris and causes damage to a third party while this insurance is in force, then such insurance should be able to provide some coverage of the space operator, unless some exclusions are applying. But in case a space object becomes a debris and the insurance has expired, and unless the operator has subscribed specific debris coverage[16], there would be no insurance

---

[16]Dedicated space debris liability coverage is available with few insurers and for low insurance amount.

in force to cover the financial liability of an operator following damages caused by its debris.

As of today, although there have been few cases of damages caused in orbit by space debris[17] there has been no application of a third party liability coverage.

With respect to damages caused on earth or in airspace[18], as per the international regulation, there is a regime of absolute liability, under which no fault has to be proved to claim for compensation. The triggering of the third party liability insurance would be facilitated in this respect. As of today, we might find on the traditional insurance market some insurance dedicated to debris, such as for debris re-entry into earth. There is very few cases (confidential ones) where the insurance has indemnified a victim due to damages caused by a space debris, but these examples aren't significant to assess any standard.

7.3.2.2.2 Third party liability insurance in the context of debris removal missions performed by a servicing satellite   The first question to be asked is who may insure these missions. In our view, it will be most likely the operator of the satellite performing the removal that will insure its mission, either because there is a legal obligation as per domestic regulation or because the operator wishes to insure itself or unless there is some agreement with the operator performing the removal.

We doubt that the operator of the satellite to be removed would actually insure the removal mission, as it would be a non-functioning satellite, unless there would be any legal obligation to do so.

The second question is what the damages are that can be covered. Basically, the insurance would afford coverage for damages caused to a third party due to the insured mission. But this insurance would not respond to damages caused by the satellite performing the removal to the satellite subject of the removal, nor in case of non-execution or improper performance of the removal service, unless the improper execution leads to a third party damage.

Therefore, we may assume that the space third party liability insurances would afford a coverage with respect to debris removal missions in the case of damages to a third party and the issue will be to know whether the insurance market is willing to offer such coverage at standard terms, conditions and price or if it would propose specific conditions and price.

Figure 7.1 shows a summary of possible insurances for space debris. This figure summarizes to what extent the existing space insurances may cover sp ace debris damages.

---

[17] Such as the collision between an operational satellite of the Iridium constellation Iridium-33 and a Russian Cosmos-2251 defunct satellite that occurred in 2009.

[18] Such as the launch of the Meridian-5 satellite on 23 December 2011 that terminated a few seconds after launch and where debris fell over the Novosibirsk Oblast in Siberia.

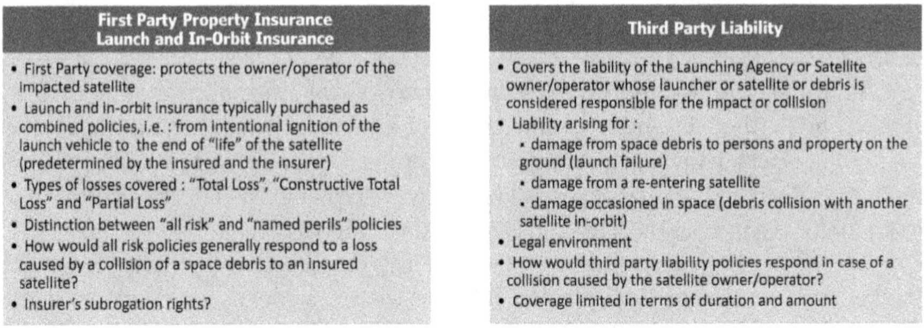

First Party Property Insurance Launch and In-Orbit Insurance	Third Party Liability

• First Party coverage: protects the owner/operator of the impacted satellite
• Launch and in-orbit insurance typically purchased as combined policies, i.e. : from intentional ignition of the launch vehicle to the end of "life" of the satellite (predetermined by the insured and the insurer)
• Types of losses covered : "Total Loss", "Constructive Total Loss" and "Partial Loss"
• Distinction between "all risk" and "named perils" policies
• How would all risk policies generally respond to a loss caused by a collision of a space debris to an insured satellite?
• Insurer's subrogation rights?
	• Covers the liability of the Launching Agency or Satellite owner/operator whose launcher or satellite or debris is considered responsible for the impact or collision
	• Liability arising for :
	  • damage from space debris to persons and property on the ground (launch failure)
	  • damage from a re-entering satellite
	  • damage occasioned in space (debris collision with another satellite in-orbit)
	• Legal environment
	• How would third party liability policies respond in case of a collision caused by the satellite owner/operator?
	• Coverage limited in terms of duration and amount

Figure 7.1: First party property and third party liability insurance contracts relevant in the space debris context

7.4 POSSIBLE EVOLUTIONS OF SPACE INSURANCE

The preceding sections have highlighted the fact that the insurance market is already affording some coverage in respect to damages caused by space debris, but with the increasing growth of space debris and development of debris removal projects, there is a room for the insurance market to offer some support in the development of such debris removal projects.

Before reviewing some paths of evolution for the insurers, a brief market review has to be done.

7.4.1 Space insurance market review

2019 was a negative year for the space insurance market, accounting for cumulative losses of less than 800,000,000 USD for a premium amounting to 400,000,000 USD (Fig. 7.2). Due to these bad results some insurers have decided to withdraw their participation in space risks[19], some others have hardened their conditions, including increase of premium rates and some others have decreased their participation.

---

[19]Andreas Berger, a Swiss Re board member in charge of the corporate insurance arm, said (...) that the company would be reducing its exposure to the space industry, as part of an overhaul of the loss-making division [11].

Premium & Claims

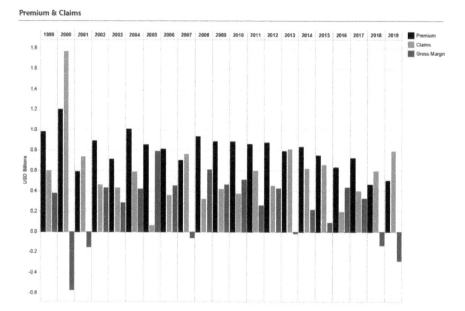

Figure 7.2: From 1999 to 2019, this figure shows the premium earned by the insurers and the claims paid for each year. The year 2019 shows clearly that the premium earned is almost twice less than the claims paid. Source: AON ISB Q1 2020 Market Report [12]

Despite these bad results, we have not seen any hardening conditions for space debris coverage, but we expect the insurers to look at more closely to the debris risks, when assessing a risk.

7.4.2 Insurance support: line of thoughts

In this section we will explore ways for the insurers to support space debris removal and even remediation projects.

Firstly, it must be noted that we cannot anticipate the reaction of the insurance market, when the exposure to space debris will become more concrete in terms of intensity of loss as well as frequency. Therefore, we believe that relying on a dedicated regulation of space debris would permit the insurers to rely on the regulation and to know what would be the extent of their exposure. A regulation would facilitate the drafting of an insurance policy specifically adapted to the regulation. The development of debris mitigation requirements, guidelines and practices is welcomed by the insurance providers as it would give better certainty for the insurance environment.

Secondly, we could explore the possibility of having a complete scheme of insurance proposed by the insurers and dedicated to debris removal missions, such as full cover including both first party property insurance and third party liability insurance. This would have the advantage of offering a full package with dedicated terms, conditions and premium rates adapted to such missions.

Thirdly, we could envision a situation where the insurers give a sort of incentive to space operators that are complying with space debris mitigation guidelines, by affording for example premium decrease (for both first party property and third party liability insurances). In this case, the insured would be somehow rewarded as acting as "good father" in complying with space debris mitigation, removal guidelines.

Fourthly, a path that could be followed is the creation of a common insurance fund, to which space operators could contribute, such a fund being fully dedicated to space debris risks. It could take the form of an international fund the aim of which will be to compensate for damage caused by unidentifiable space debris, whether occurring on Earth or in outer space. Some experts have brought up the idea of having the fund contributed to by the space community, by way for example of an obligation to participate in it for each object to be launched, a contribution that could also be based on the debris mitigation plans to be followed by the participating space operator[20].

Fifthly, aside from the application of a strict census of insurance coverages, there is a legal and contractual system that could limit the exposure of the insurers and therefore, reduce the insurance premium. Such a system is commonly known under the appellation of cross-waiver of liability and hold harmless. The use of said clauses within a contractual relation leads to limiting the liability that is borne by a space operator vis-a-vis its contractor and third party by having its liability undertaken totally or partially by its contractor. The contractual practices between space actors may benefit insurers as well in limiting their exposure to the extent that the liability of the insured is limited.

Finally, there is another protection of a space operator to be explored, which is the application of State liability above the liability cap of an authorized operator as set up by some domestic space regulations[21]. Under these regulations, the authorisation State agrees to bear the liability of its authorised operator above a certain liability cap (sometimes associated with an insurance obligation). In this case, the space operator doesn't bear any liability above the cap and will rely on the State to indemnify the eventual victims. We could think of extending this mechanism to damages caused by space debris (above or not above a certain liability cap) and having the State liable for this type of risk. As a complement and to permit the State not to be fully exposed, an insurance fund could be put in place, in order to cover the liability of the State.

7.5 CONCLUSION

Space debris today is a growing risk exposing the operators to seeing their liability eventually put at stake. Even though there has been very limited examples of damages caused by space debris, either on earth or in-orbit and even though the actual space

---

[20] See SPACE DEBRIS AND THE LAW - Dr. Frans G. von der Dunk [13].

[21] Such as French, US and UK to some extent. France, Loi n° 2008-518 du 3 juin 2008 relative aux opérations spatiales [14], UK, Space Industry Act 2018 [15], United States of America, U.S Commercial Space Launch Competitiveness Act

regime does not provide much room for claims in the case of in-orbit collision, the risk is real. The multiplication of space debris automatically increases the risk of damages caused by debris.

Today, space debris risk is partially transferred to the insurance market, thus protecting the space industry. But we believe that enacting a dedicated regulation to space debris would provide some comfort to the insurers in the legal environment surrounding debris liability and would enable them to offer adapted insurance solutions based on the applicable regulation.

The space insurance community is today affording some coverage solutions in respect to damages caused by space debris, but more adapted solutions could be thought of in order to offer better and greater protection in respect to debris mitigation or removal projects.

ACRONYM

CNES French Space Agency Centre National d'Etudes Spatiales

COSPAR Committee on Space Research

ESA European Space Agency

EXA Ecuador's Space Agency Agencia Espacial Civil Ecuatoriana

IADC Inter Agency Space Debris Coordination Committee

NASA National Aeronautics and Space Administration

NASDA National Space Development Agency of Japan

COPUOS Committee on the Peaceful Uses of Outer Space

UNOOSA United Nations Office for Outer Space Affairs

GLOSSARY

Absolute liability: refers to the liability system which does not require a proof of fault

Space object: space objects operated partially or totally in-orbit

All risk: refers to the insurance policies based on an inclusion of all risks except the one specified in the policy

Authorization: refers to an act given by an authority allowing a space activity

Cross-waiver: clause under which a contractor agrees not make a claim against its co-contractor

Custom: source of international law created by state's practice

Debris removal missions: space activity in order to remove space debris by moving the debris to a cemetery orbit or back into the atmosphere

Due diligence: refers to the necessary measures taken by a state to avoid committing a tort

Fault-based liability: refers to the liability system which requires a proof of damage, fault and causation link

First party property insurance: insurance covering an insured asset in case of loss of said asset

Hold Harmless: clause under which a contractor undertakes to bear partly or totally the consequences of the liability of its co-contractor

In-orbit insurance: first party property insurance after launch phase and during life-time in orbit

International wrongful act: refers to the violation of an international obligation

Launch insurance: first party property insurance during launch phase

Liability Convention: Convention on International Liability for Damages Caused by Space Objects

New space: refers to the new space industry from the private-sector initiative

Outer Space Treaty: Treaty on Principles Governing the Activities of States in the Exploration and Use of Outer Space, including the Moon and Other Celestial Bodies

Risk assessment: analysis of risks existing in space activities

Space insurance market: the community of insurers offering space insurances

Third party liability insurance: insurance covering the consequences of damages caused to third party and due to an insured activity

FURTHER READING

Hall, G. E. (2007). *Space debris—an insurance perspective.* Proceedings of the Institution of Mechanical Engineers, Part G: Journal of Aerospace Engineering 221.6 : 915–924.

Chrystal, P., McKnight, D., Meredith, P., Schmidt, J., Fok, M., Wetton, C. (2011). Space debris: On collision course for insurers?. *Swiss Reinsurance Co. Publ.*, Zurich, Switzerland.

Masson-Zwaan, T., Hofmann, M. (2019). *Introduction to Space Law.* 4th Edition, Wolters Kluwer.

Smirnov, N. N. (2001). *Space Debris: Hazard Evaluation and Debris*, Earth Space Institute Book Series 6 (English Edition). CRC Press.

Bibliography

[1] Secretariat of the United Nations. *Treaty on Principles Governing the Activities of States in the Exploration and Use of Outer Space, including the Moon and Other Celestial Bodies, 10 October 1967*, volume 610. United Nations Treaty Series (UNTS), 1967.

[2] Secretariat of the United Nations. *Convention on International Liability for Damage Caused by Space Objects, 1 September 1972*, volume 961. United Nations Treaty Series (UNTS), 1972.

[3] United Nations Office For Outer Space Affairs (UNOOSA). *Space Debris Mitigation Guidelines of the Committee on the Peaceful Uses of Outer Space*, volume 9-88517. United Nations, 2010.

[4] Secretariat of the United Nations. *Treaty on Principles Governing the Activities of States in the Exploration and Use of Outer Space, including the Moon and Other Celestial Bodies, opened for signature in Washington, London and Moscow, 27 January 1967*, volume 610. United Nations Treaty Series (UNTS), 1967.

[5] International Law Commission (ILC). *Draft Articles on the Responsibility of States for Internationally Wrongful Acts, Report of the ILC on the Work of its Fifty-third Session, UN GAOR, 56th Sess, Supp No 10*, volume UN Doc A/56/10. United Nations, 2001.

[6] Stephan Hobe, Bernhard Schmidt-Tedd, Kai-Uwe Schrogl, and Peter Stubbe. *Cologne Commentary on Space Law: Rescue Agreement, Liability Convention, Registration Convention, Moon Agreement*, volume 2. 2013.

[7] H. Ijaiya. Space debris: Legal and policy implications. *Environmental Pollution and Protection*, 2(1):25–26, 2017.

[8] European Space Agency. "Space Debris by the Numbers". Information correct as of January 2019, (accessed in December 2019). . https://www.esa.int/Safety_Security/Space_Debris/Space_debris_by_the_numbers.

[9] I. Caselli. "Ecuador Pegasus satellite fears over space debris crash". BBC News (24 May 2013): . http://www.bbc.com/news/world-latin-america-22635671.

[10] The San Diego Union-Tribune. "Ecuador writes off ill-fated satellite". the San Diego Union-Tribune (6 September 2013). https://www.sandiegouniontribune.com/en-espanol/sdhoy-ecuador-writes-off-ill-fated-satellite-2013sep06.-story.html.

[11] C. Cohn and T. Sims. "Space insurance costs to rocket after satellite crash". Reuters (31 July 2019). https://fr.reuters.com/article/companyNews/idUKL8N24V2MT.

[12] International Space Brokers. AON ISB Q1 2020 Market Report. https://www.aon.com/industry-expertise/space.jsp, 2020.

[13] Frans von der Dunk. Space debris and the law. *Proceedings of the 3rd European Conference on Space Debris, ESOC, Darmstadt, Germany, 19 - 21 March 2001 (ESA SP-473, August 2001)*, 2001.

[14] Secrétariat général du gouvernement (SGG). "LOI n° 2008-518 du 3 juin 2008 relative aux opérations spatiales (1)". NOR: ESRX0700048L, Légifrance.gouv.fr. `https://www.legifrance.gouv.fr/affichTexte.do?cidTexte=JORFTEXT000018931380`.

[15] The National Archives on Behalf of HM Government. "Space Industry Act 2018". `http://www.legislation.gov.uk/ukpga/2018/5/contents/enacted`.

VI

Risk Complexity and Space Debris Governance

CHAPTER 8

Space Sector Resilience and Ways to its Governance

Olga Sokolova

Sirin Orbital Systems AG, Zürich, Switzerland
Paul Scherrer Institut (PSI), Villigen, Switzerland

Matteo Madi

Sirin Orbital Systems AG, Zürich, Switzerland

CONTENTS

ISTORICALLY, the public awareness of space hazards was limited to the countries, who had operational scientific space missions. Risk assessment methods for technological systems were traditionally based on vulnerability identification for a specific component to an adverse event followed by its functional loss. The subsequent risk management was focused on hardening specific components to an acceptable risk level in order to prevent the overall system failure. The New Space sector's rapid development challenges the resilience concept for this sector. Moreover, the ongoing growing volume and variety of data and signals affect the reliable operation of down-stream (ground-based) systems. The "space paradigm shift" results in an increasing number of uncertainties and interdependencies. This chapter questions relevant stakeholders on resilient spaceborne infrastructure assessment, describing methodologies for its assessment and adapting quantitative indices, and how the space debris risk can be governed. The scenarios of inter-infrastructural failures are also presented in this chapter. The assessment of potential loss increases the awareness of the relevant stakeholders of the importance of the efforts dedicated to mitigate the risk. The detailed studies have shown society that the "zero-risk-level" does not exist. Peril assessment is important for designing mitigation solutions, since the evaluation of risk provides a basis for planning and allocation of limited resources: technical, financial and others.

8.1 SPACEBORNE ASSETS AS A CRITICAL INFRASTRUCTURE. INTER-INFRASTRUCTURAL CATASTROPHE SCENARIO

Through the years, the vertically integrated ground-based systems, with only a few points of communication, turned into complex horizontally integrated systems, with many points of interaction in many of their dimensions [1]. New Space development ravels the picture even more. The growth of key business players and technological capabilities increases the system's complexity for modelling, analysis and operation. The New Space business model shift drives higher dependence of ground-based infrastructures on reliable and high quality space asset operation for a variety of purposes. In turn, this expands the number of interdependencies between them.

Almost every critical infrastructure relies on space assets' reliable operation. However, the definition of critical infrastructure has changed over the time and differs from country to country. The first formal definition of "critical infrastructure" was given in Executive Order 13010 [2] and described as so vital that their incapacity or destruction would have a debilitating impact on defence or economic security. It was later re-defined in the Patriot Act of 2001 [3] as systems and assets, whether physical or virtual, so vital to the United States that the incapacity or destruction of such systems and assets would have a debilitating impact on security, national economic security, national public health or safety, or any combination of those matters. Though the number of sectors has changed over the time, the definition remained the same. The Organization for Economic Cooperation and Development (OECD) defines the critical infrastructure as an infrastructure that provides an essential support for economic and social well-being, public safety, and the functioning key government responsibilities, such that disruption or destruction of the infrastructure would result in catastrophic and far-reaching damage [4]. European Union Commission set an overall policy approach and framework for critical infrastructure protection and specified critical infrastructure as physical and information technology facilities, networks, services and assets that, if disrupted or destroyed, would have a serious impact on the health, safety, security or economic well-being of citizens or the effective functioning of governments in EU States.

Almost all spacefaring nations have legally defined critical infrastructure and made a list of corresponding sectors. Among active players with full launch capabilities only Russia does not have a relevant definition. However, such a definition was proposed by Sokolova, O. and Popov, V. [5] on the basis of "National Protection

Strategy" adopted on 12th May 2009, and "List of Perspective Areas of Russian Economy and Critical Technologies (Order no.899 from 7th July 2011)", "List of Backbone Enterprises from 8th February 2015". Though the definitions for critical infrastructures look similar, certain differences can be noticed. Currently, the space industry is listed in the European list of critical infrastructures together with ten other sectors [6]. The role of the space sector as a critical infrastructure is rapidly changing from pure scientific missions to being an active player in the future economy. It appears that the role of an infrastructure transforms over time, and systems previously treated as non-critical are nowadays identified as critical. Therefore, it is proposed to consider the space industry as a typical example of a critical infrastructure; if it cannot, at least in short term, be substituted by another system, it results in systematic failures across several infrastructures that have been created in an interconnected way for increasing their efficiency.

Four critical infrastructure states are determined: operable, under threat, vulnerable, and inoperable. Modern infrastructure networks do not operate in isolation, so they are interdependent, meaning the failure can propagate between the sectors. Both structural and dynamic complexity can be observed. Heterogeneity is a common feature of a structural complexity, which refers to the differences in the elements, their interconnections and roles within the system structure, often with high-connected core elements and low-connected periphery nodes [7]. For instance, the power grid is a typical example of a heterogeneous network. However, heterogeneity is not dominant in modern space systems, though geostationary orbit (GEO) telecommunication satellites can be treated as those with higher hierarchy as is shown further. The features of dynamic complexity are: self-organization, emergent behaviour and adaptation. Self-organization stands for a specific dynamic feature of a complex system, which amounts to the capability of re-organizing its isolated elements and subsystems into coherent patterns without intervention from external or a central authority [8]. A system exhibits emergence when there is a coherent emergent at the macro-level that dynamically arises from the interactions between the parts at the micro-level [9]. There are four schools of thought that study emergence, as summarized by De Wolf, T. and Holvoet, T. [10]. Adaptation relates to the ability of adjusting the system structure and behaviour to respond to external pressures using long-term memory experience. These features were considered in Section 8.3 while submitting the corollaries for space system resilience analysis.

Several types of interdependency are distinguished [11]:

- *Physical* – two infrastructures are physically interdependent if the state of each is dependent on the material output(s) of the other.

- *Geographic* – infrastructures are geographically interdependent if a local environmental event can create state changes in all of them.

- *Cyber* – an infrastructure has a cyber interdependency if its state depends on information transmitted through the information infrastructure.

- *Logical* – two infrastructures are logically interdependent if the state of each depends on the state of the other via a mechanism that is not a physical, cyber, or geographic connection.

- *Policy* – the change in form of a policy/procedure that takes effect in one of the systems influences the other.

- *Societal* – infrastructure operation is affected by the public opinion.

It is clear that the last four types of interdependencies are relevant for the studied problem. Though it may sound like geographical interdependency is also relevant, it is not the case. Space weather is a hazard, which is a relevant case. Even though a space weather hazard may simultaneously impact both space assets and ground-based critical infrastructures, the physical mechanisms differ. The space weather impact on satellites results in increasing microelectronics upset rates and creating electrostatic charging hazards [12]. The severity of the impact depends on the space weather manifestation type and the asset characteristics. The National Oceanic and Atmospheric Administration (NOAA) introduced a system of indices for describing solar radiation storms (**S** index) and geomagnetic storms (**G** index), scaled from 1 (minor) to 5 (extreme). The space weather manifestation, which is hazardous for satellites, can pose no, or a minimum threat, to ground-based critical infrastructures. It is believed that satellites are more likely to be at risk from fast solar wind stream events than a Carrington-type storm caused by coronal mass ejection [13], while a Carrington-type event is treated is the most probable catastrophe scenario for modern power grids. Authors exclude geographic infrastructure interdependency in the further analysis.

When systems are rafted together, the critical elements of each become critical elements of all, because of the possibility that failure in one part of one system will be externalized to others [14]. Rinaldi, S. [11] also specifies failure types. They include cascading failures, when the failure in one infrastructure causes a disturbance in another infrastructure; escalating failures, when failure in one infrastructure worsens an interdependent disturbance in another infrastructure; and common cause failure, when two or more infrastructures are disrupted at the same time due to a common cause. It is difficult to anticipate the situations when simultaneous failures bring into play dormant and previously hidden interdependency pathways which destructively and synergistically amplify the failure [15]. There is a trend to use modelling and simulation techniques to reveal infrastructure interdependencies such as network topology [16], graph theory [17], Multi-area Thevenin Equivalent [18], and others. Interdependencies can be unidirectional or bidirectional, when a component depends on another through some links, and this latter component likewise depends on the former component through the same and/or other links [7].

The analysis of inter-infrastructural catastrophe is important for addressing, (1) the processes and consequences of physical infrastructure failures in terms of physical capital losses and service flow disruptions, and (2) the resulting business disruptions and economic flow losses across the wider macroeconomic sectors [19]. Observations of inter-infrastructural catastrophe are rare and incomplete. Some countries may have developed detailed models, but they are not in the open access.

Nearly 2,000 satellites operated by 60 countries and 20 organizations orbited the Earth in February 2016, and the number grew by 700 objects in the last 4 years. The satellite mass dropped down to 200 kg and even less compared to the typical mass of 500 to 2,000 kg in the Old Space era. The change also happened in the launching facilities landscape. In addition to a few established launch sites, new launch systems and ride-share options have been developed. Depending on the mission goal, they can be placed in various orbits named Low-Earth Orbit (LEO), Medium-Earth Orbit (MEO), Geostationary Orbit (GEO), Polar, Molniya, Tundra, Sun-synchronous Orbits, or on Lagrange points. Figure 8.1 shows most common orbit altitudes for LEO, MEO, and GEO.

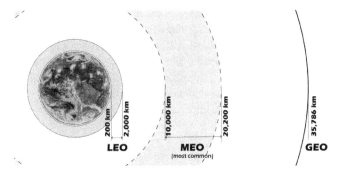

Figure 8.1: A classification of most common Earth orbits by altitude

For instance, placing a satellite at GEO gives pretty large coverage area per satellite, which can be achieved only by placing a constellation of satellites on MEO. Yet the space infrastructure consists not only of satellites. The whole spectrum of space assets includes:

- **Global Navigation Satellite System (GNSS)** – representing an asset which is among the most critical ones for a wider range of ground-based critical infrastructures.

- **Communication Satellites (COMM)** – whose impact on modern economy and society's well-being is staggering. They are precious for the proper functioning of ground-based critical infrastructures reliant on robust communication.

- **Earth Observation Systems** – including remote sensing and meteorological satellites (METEO) that routinely monitor and provide early warning for environmental threats and extreme weather patterns.

- **Space Probes** – which are for exploratory missions and located at the vicinity of the Earth. One can conclude that they have minor impacts on ground-based critical infrastructures. However, certain space probes are responsible for space weather forecasting.

- **Launchers and Space Stations** – do not directly impact ground-based critical infrastructures.

- **Miniaturized-Satellites** – which can carry out various functions. They are at their infancy, albeit their affordability in terms of fabrication know-how, accessibility and relatively low cost, make them a booming solution for many applications including the above mentioned GNSS, COMM and METEO. These small-satellites are the backbone of the New Space era.

Almost every ground-based critical infrastructure relies on space assets' reliable operation. The inter-infrastructure failure scenarios are presented in Fig. 8.2. The patterns reflect different critical infrastructure operational states. In case a sector's pattern is blank, it is meant for the situation when the infrastructures are predominantly marginally dependent. Three scenarios are represented in the case of GNSS, communication or meteorological satellites loss. All of them correspond to the worst-case scenario triggered by one of the hazards. The number of small-satellites is rapidly growing as well as the number of functions they carry. Currently, ground-based critical infrastructures do not have strong dependencies on them. However, the situation is swiftly changing with the appearance of satellite mega-constellations.

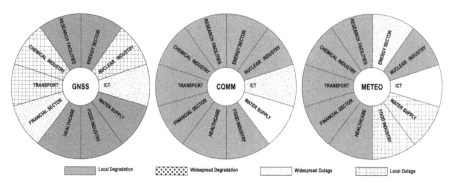

Figure 8.2: Ground-based critical infrastructure disruptions caused by the multiple space assets loss

Ground stations form another group of space assets, which are in fact ground-based infrastructures. They monitor and control the health and status of space assets, send commands and receive data. The consideration of ground stations as space assets extends the list of hazards by amending ground-based technological and natural perils. Nonetheless, redundant capabilities, i.e. alternate control centres, are normally established. Therefore, critical infrastructure degradation scenarios presented in Fig. 8.2 remain unchanged.

A number of studies suggest that space asset loss can have significant economic impacts on other sectors. Weather forecasting is a good example, which has far reaching impacts across several industries. The understanding, coupled with data from several satellites, has led to an improved ability to predict the return of El Niño, which can then be used to alert weather-sensitive industries around the world that they may face increased risk of experiencing abnormal weather phenomena in their regions [20]. Accurate weather forecasts are crucial in maintaining the reliability of the supply of electricity to users through management of the power grid (especially, close monitoring of overload conditions). The benefit values of more precise ground-based weather forecast facilities are significant, often reaching millions of US dollars in the electric utility industry [21].

In natural disasters management, the benefits from space infrastructure are clearer. Space-based observations allow the Earth to be seen as a dynamic integrated system of land, water, atmosphere, ice and biological processes, while satellite telecommunications allow worldwide connections [22]. Extreme weather and climate events are among the primary causes of infrastructure damage causing large-scale cascading outages, or shifts in the end-use electricity demands leading to supply inadequacy risks [23]. Analysis of the number of events per year and associated losses in the past 40 years showed the rising trend, and the ranking of risks "extreme weather events" and "major natural disasters" is constantly increasing over the years in the risk matrix developed by World Economic Forum (WEF). Global risks are not strictly comparable across years, as definitions and the set of global risks have evolved with the new issues emerging on the 10-year horizon [24].

The impact of GNSS loss for UK can be taken as a case-study. In the "GNSS-reliant present-day UK", the economic impact of the GNSS loss has been estimated at 5.2 bn pound sterling over a five-day period, comprising 1.7 bn in lost Gross Value-Added (GVA) benefits and 3.5 bn in lost utility benefits [25]. The three most affected sectors are ground transportation, emergency and justice services and maritime transportation. For instance, maritime transportation would face disruption in all routine ports and the loading and unloading of containers for the duration of the outage. In total they account for 67% of all impacts.

Generally, the financial cost associated with system loss can be divided into three categories:

- **Direct costs**, the calculation of which provides insight into the sales revenue experienced by businesses caused by the impact itself. The first step in direct cost assessment is "impact fields" identification. To elicit the direct costs and benefits for each of the impact fields, the following types of costs can be distinguished: (1) changes in production cost structure (changed input structure or technology); (2) change in productivity (changed output for the same amount of input); (3) change in final demand (demand shift as response to lower supply); (4) replacement costs (investment to reinstall damaged assets and infrastructure); (5) preventive expenditures (for replacing one service with another system); (6) change in public expenditure (transfers and subsidies, e.g. for damage compensation) [26]. For instance, the impact fields can be characterized by national input-output tables often classified by NACE codes (Nomenclature des Activités Économiques dans la Communauté Européenne) as: a single NACE sector represents an impact field; several distinct NACE sectors represent an impact field; a new sector is extracted from several distinct NACE sectors; a single NACE sector is disaggregated into two or more subsectors.

- **Indirect costs**, which is the sector's indirect performance degradation due to: the reduction of the value added to the economy by the sector itself (profit, tax, etc.); the reduction in the sector's immediate consumption and demand for services and goods by other sectors; and the reduction of sectors liability to meet demands of existing customers. In other words, both upstream and downstream indirect shocks are considered to propagate along the tier after the tier of suppliers and customers respectively, which collectively constitute the supply chain of the economy [27]. The indirect losses value can vary within the interval limited by the losses caused by the production loss indirectly caused by other direct shocks (minimum bound) and loss in certain facilities who do not experience power shortage by themselves (upper bound). There are still no consequences in the way to calculate indirect or high-order economic losses.

- **Resulting long-term costs** of macroeconomic relevance. The three main approaches are: econometric models, which use statistical methods to analyse past time series on the economic performance of a region in order to project economic activity into the future; input-output analysis based on the input-output tables of economies that depict interdependencies between economic sectors; and computable general equilibrium modelling, which goes beyond input-output models by linking industries via economy-wide constraints including constraints on the size of government budget deficits; constraints on deficits in the balance of trade; constraints on the availability of labour, capital, and land; and constraints arising from environmental considerations, such as air and water quality, etc. [28][1].

There is still no consensus on the correct approach for estimating economic loss. The method commonly known as input-output analysis has gained the most attention in recent years for its ability to model indirect or higher-order economic losses. The input-output approach spawns an entire field of related models and includes the inoperability input-output model, the Ghosh supply-side model, the dynamic input- output models, the key-linkages analysis, as well as the inventory based models amongst others [30]. World Input-Output Database (WIOD) is a unique data source which provides underlying data covering 43 countries and 56 economic sectors [31]. It has high transparency over underlying data sources and methodologies used. One inevitable drawback in this type of analysis stems from the lack of precise space statistics [22].

The space asset loss may overall have global impacts, and the impact can go beyond the borders of a certain country who for instance owns the satellite. One can compare effects with those observed during the pipeline loss, which carries oil/gas belonging to a country far away from the mining place, albeit this case mainly reflects physical interdependency. This fact drives awareness growth even in non-spacefaring nations. Nevertheless, industries, whose operation does not rely on space systems, are less concerned even in spacefaring nations. Several factors define the economic loss from a catastrophe. Countries with higher per capita income experience a similar number of catastrophic events but suffer less death from these events [32]. Risk averse individuals will make different risk-return trade-off choices at different income

---

[1]The examples of long-run economic performance assessment after a shock is presented by Akao, K. and Sakamoto, H. [29].

levels [33]. The latter reference identifies a number of social critical factors: nations with higher levels of educational attainment and greater openness for trade are less vulnerable to disasters; stronger financial sector and a smaller size of government (measured as the fraction of government expenditure per GDP) are associated with a lower disaster death toll. Institutional quality and international openness mitigate negative consequences [34]. Moreover, the poorer countries are also unlikely to be able to adopt the counter-hazard policies [35].

In addition, businesses will incur indirect costs through the need to establish operational procedures to monitor and initiate adaptation measures when needed. Two types of costs are distinguished: *ex ante* – mitigating the catastrophe, and *ex post* – coping with the consequences [36]. Nevertheless, the economic benefits of mitigation could exceed the ex post cost [37]. It is important to note that stakeholders should not be just aware of a catastrophic event. Vulnerability may arise due to continuing degradation as a consequence of many smaller impacts [38].

8.2 RESILIENCE DEFINITIONS AND METRICS IN VIEW OF SPACE-BORNE INFRASTRUCTURE

Space systems are designed for functioning in the most hostile environment among known ones. A brief overview of hazards to space assets is given by Sokolova, O. and Madi, M. [39]. Nevertheless, the marginal productivity constraint is the main factor which determines characteristics of New Space assets. It means that the function is provided with the minimum number of required satellites. Therefore, each satellite becomes a highly vulnerable asset. This fact even heightens its cruciality. In general, the reliability requirement of an asset is determined by the degree of its influence on the system's security. However, the uniqueness of each mission complicates uniform reliability protocol implementation. The total mass, composition material and construction procedure limitations, driven by the cost-reliability criteria, are unique for each mission. The common understanding of the space debris threat is the first step in improving this sector's resilience.

The majority of risk assessments take the approach specified in ISO 31000 [40]. This standard promotes the idea of risk identification as the first step of risk management. Overall, the risk management paradigm consists of five interconnected phases named analysis, evaluation, research/control, communication, and monitoring. The risks and uncertainties determination is done in the first phase, which are later identified at the evaluation phase. Phase two is also devoted to assessing risks and uncertainty reduction options. The third phase consists of two parallel actions such as research, which targets uncertainties reduction and control, i.e. reducing and controlling the risks. The risks and mitigation management are delivered to the stakeholders within the fourth phase. The ongoing confirmation/revision of assumptions about the risks is made in the last phase. A graphical representation of the paradigm is given in Fig. 8.3.

In terms of infrastructure assessment, two corollaries should be accepted: (1) one should articulate the studied system with a well-defined purpose along with relevant stakeholders goals [41], (2) one should study the system and its environment together,

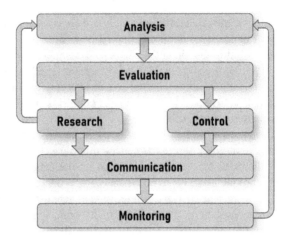

Figure 8.3: Risk management paradigm

where environment is actually anything outside of the direct control of the system and includes any phenomenon influencing the processes and behaviours of the system [42]. The National Infrastructure Protection Plan (NIPP) developed a framework for risk mitigation, which is rather a construct for risk management than a perspective approach. It was originally introduced in 2006 and modified in 2013. The framework is presented in Tab. 8.1.

Table 8.1: Risk management framework

Phase	Related action
Set goals & objectives	• Set broad goals for infrastructure security and resilience • Determine collective actions through joint planning efforts
Identify infrastructure	• Analyse associated dependencies and interdependencies[2]
Assess & analyse risk	• Improve information sharing • Apply knowledge to enable risk-informed decision making
Implement risk management activities	• Rapidly identify, assess, and respond to cascading effects during and following incidents • Promote infrastructure, community, and regional recovery following incidents
Measure effectiveness	• Learn and adapt during and after exercises and incidents

Resilience is a key strategy for handling risk as we study current risk management and governance frameworks [44]. The basic reason for this shift was the extension of the traditional system limits. It is shown that adding the resilience concept to the modelling procedure gives more realistic results. However, a separation between the schools, once centred around the risk, the other around resilience is observed, especially in engineering environments [45]. Risk to a certain system can be described as in Eq. (8.1):

$$
\begin{aligned}
Risk &= (A, C, U) \\
&= (A, U) + (C, U \,|\, A) \\
&= occurrence\ of\ events,\ and\ associated\ uncertainties \\
&\quad + consequences\ of\ given\ events,\ and\ associated\ uncertainties
\end{aligned}
\tag{8.1}
$$

where A represents events (hazards, threats, changes); C is the consequences; U is the uncertainties. The symbol "+" should not to be interpreted as a mathematical sum, but as a symbol for combining two elements [45].

The common approach includes the process of a series of individual risks identification, assessing their likelihoods and consequences followed by the comparison of risks based on the severity of impacts. The analysis results are later used for risk matrix construction following an intervention *as low as reasonably achievable*. In contrast, the later developed "vulnerability" concept manages the infrastructures following an idea *as resilient as reasonably achievable*. Vulnerability is a degree to which a system, subsystem, or system component is likely to experience harm due to exposure to a hazard, either a perturbation or a stress/stressor. The difference between the two concepts is visualized in Fig. 8.4. While the consequences of a certain hazard on system properties is studied in a risk concept, the impact of a variety of stressors is examined in a vulnerability concept.

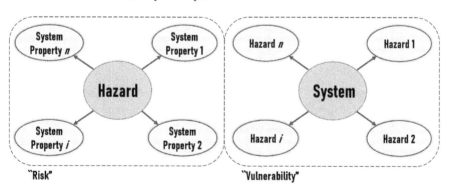

Figure 8.4: Difference between risk assessment and vulnerability assessment approaches

The disadvantage of using the risk concept is that it fails to assess non-linear/complex risks [46]. Cause-effect relationships for complex risks can be understood only in hindsight after a disaster occurs. It differs from linear/complicated risks for which the cause-effect relationship can be understood in advance and stochastic approaches are appropriate for their assessment. It is of particular importance for the space debris risk.

The criticism of over-protected system design, construction and maintenance triggered the resilience concept popularization. It is known that a small increase in the protection level may require a large amount of additional costs. In other words, achieving desired a protection level is normally not cost-effective in relation to the actual hazards. In contrast to risk assessment, which measures a potential loss associated with certain uncertainties, resilience is a much wider concept. Resilience (or resiliency) originates from the Latin word *resilio* which literally means "to jump back" [47]. In 1973, C. S. Holling [48] defined resiliency as a measure of the persistence of systems and of their ability to absorb change and disturbance and still maintain the same relationships between populations or state variables. After Holling, numerous interpretations of resilience have been developed. A summary of resilience definitions and key properties of each definition is given by Francis, R. and Bekera, B. [49]. However, a general definition of resilience is given by the United Nations Office for Disaster Risk Reduction (UNDRR). According to the UNDRR, resilience is the ability of a system, community or society exposed to hazards to resist, absorb, accommodate, adapt to, transform and recover from the effects of a hazard in a timely and efficient manner, including through the preservation and restoration of its essential basic structures and functions through risk management [50]. UK Cabinet Office [51] emphasizes that a resilient system is able to absorb lessons for adapting its operation and structure to prevent or mitigate the impact of similar events in the future.

A resilient system is specified as a system of **four "R's"**:

- *Robustness* – strength, or the ability of elements, systems, and other measures of analysis to withstand a given level of stress or demand, without suffering degradation or loss of function. It also includes robust human resources. It refers to the functionality right after the extreme event. Satisfying this parameter targets correction of design issues as poorly detailed, improperly restrained or vulnerable.

- *Redundancy* – capacity of satisfying functional requirements in the event of disruption, degradation or loss of functionality. It refers to other measures or systems which are substitutional to existing ones, as described in Eq. (8.2).

$$Redundancy = f \left(reserve\ capacity, \frac{1}{time\ to\ access} \right) \qquad (8.2)$$

- *Rapidity* – the capacity to meet priorities and achieve goals in a timely manner in order to contain losses, recover functionality and avoid future disruption. It represents the slope of the resilience curve, Eq. (8.3), during the recovery phase. In addition, it shows how fast the society can learn from the event.

$$Rapidity = \frac{dQ(t)}{dt} \qquad (8.3)$$

where $Q(t)$ is a functionality level.

- *Resourcefulness* – the capacity to identify problems, establish priorities, and mobilize alternative external resources when conditions exist that threaten to disrupt some elements, system, or other measures. It is sometimes thought of as how to improve redundancy by providing measures to maintain additional resources and increase rapidity *ex post* by making investment *ex ante*.

The conceptual resilience curve associated with an event is represented by Panteli, M. [52]. Three levels of resilience are mainly used[3]. The level R_0 stands for the initial level. A system can be considered as a resilient one if the initial resiliency level R_0 is high enough to withstand the extreme events. Values R_{pe} and R_{pr} respectively indicate the post-event resiliency level and the post-restoration resiliency level. The level R_{pr} can be the same as R_0 ($R_{pr} = R_0$), higher ($R_{pr} > R_0$) or lower ($R_{pr} < R_0$). The post restoration state is ended at the moment t_{pir}. The features of the system such as resourcefulness, redundancy and adaptive self-organization not only define the post-restoration resiliency level R_{pr}, but also the time frame of restorative state $t_r < t < t_{pr}$. The mitigation actions should aim at reducing the drag in resiliency levels after the event ($\Delta R_0 - R_{pe}$) and reducing the recovery time ($\Delta t_{pr} - t_r$).

Industry distinguishes several types of resiliency. "Operational resiliency" refers to the characteristics which would help a system to maintain operational strength and robustness in the face of an extreme event. "Infrastructure resiliency" refers to the physical strength of a system for minimizing the portion of the system that is damaged, collapsed or in general becomes non-functional [53].

In accordance with the risk type, two resilience building approaches can be distinguished: specified and general. The specified resilience refers to the known risks, whose effects have already been observed in the past, and risk assessments are based on linear cause-effect relationships [54]. The WEF describes strategies for building the specified resilience [55]. Case studies were elaborated for building specified resilience to particular shocks [56]. The threats to New Space assets are outside the scope of experience. Thereby, general resilience, which is an ability to withstand unknown shocks [57], covers better the needs of the New Space sector. The challenge of building resilience to unknown disturbances is far more difficult than planning for known types of disturbances, and like any management strategy it comes with a cost [58]. Approaches to build general resilience have to be simultaneously implemented top-down and bottom-up.

Overall, resilience frameworks target either system assessment or system enhancement. In the assessment process, the system is evaluated with respect to given threats and the system's properties, the four "R's". The system's enhancement approach is aimed at defining the proportion between satisfactory parameters values, cost-benefit analysis and stakeholder engagement. The enhancement strategy proceed as an operation-based resilience or planning-based resilience strategy.

As it was mentioned earlier, one of the main challenges is the limitation and

---

[3]See figure tagged, "A conceptual resilience curve associated with an event.", in Panteli, M. [52].

the scarcity of historical data. Despite the fact that the impact of threats on space assets was observed in the past, the revolutionary change in operation procedures and performance assessment algorithms make it difficult to correlate the events. History teaches us that:

> [extreme] events of the past may not lead to extreme consequences in the present and vice versa.

It is proposed to develop a common New Space resilience assessment methodology using the World Bank's recommendations for managing environmental disasters [59], as follow:

- make information on disaster risk easier to access;
- take preventive measures;
- provide adequate infrastructure and public services to reduce vulnerabilities;
- build institutions that permit public oversight of disaster preparedness and disaster response.

The choice of metrics for measuring resilience is done by answering questions: "resilience of what?", "to what?", "for whom?", and "for what purpose?". In this case metrics provide a solid basis for monitoring, evaluating, reporting and decision-making. The examples of further precision of the question-set is given in Tab. 8.2.

New Space sector resilience is an inter-disciplinary problem. The variables for evaluating resilience should be taken from the following fields: natural, engineering, social, economic and institutional. Similar to [60], a scoring matrix to evaluate the system's capability to plan, absorb, recover and adapt can be developed. The choice is supported by the domains of resilience defined by Hosseini, S. [61], such as organizational, societal, economic, engineering. Consequently, the method calibration for enhancing resilience should be performed in respect to the set of indicators that cover the following problems: various stakeholders' interest consideration, intervention in resilience development phases (it can be applied in one or more phases), threat specification, interdependencies reflection, and socio-economic behaviour assessment.

Our knowledge of the complex adaptive system (e.g. space infrastructure) is always partial and incomplete due to its dynamic nature. Processes and system's attributes are constantly changing over time. One should evaluate to which level chosen resilience metrics meet the needs for which they were designed in order to be effective. In the first step, New Space sector resilience can be expressed with the given common set of matrix. The system's capacity R_C can be described as in Eq. (8.4):

$$R_C = \frac{Q\left(t_D\right)}{Q\left(t_E\right)} \times \frac{Q\left(t_R\right)}{Q\left(t_E\right)} \times S_P \tag{8.4}$$

where $Q\left(t_R\right)$ is a restored condition, $Q\left(t_D\right)$ is a damaged condition, $Q\left(t_E\right)$ is the normal condition, and S_P is a recovery speed defined as $S_P = \left(t_D - t_E\right) / \left(t_R - t_E\right)$. The terms t_R, t_D and t_E are respectively the time when restoration was completed, time of the worst damaged state, and time when the event started.

Table 8.2: Specification of performance variables

Category	Performance variable
Resilience *of what?*	
Function	System function, output service, requirements, capacity, ability
State	System state, equilibrium
Structure	System structure, components, relationships, feedbacks, connectedness, connectivity level
Degradation	System degradation, vulnerability, damage
Loss	Loss minimization, destruction, outages
Growth	Growth, growth trajectory/ path
Resilience *to what?*	
Disruption	Disruption, interruption, disturbance, accident, perturbation
Change	Change, shift, discontinuity
Event	Event, incident, occurrence
Damage	Damage, disaster, emergency, negative impacts, accidents
Failure	Failure, fault, breakdown
Uncertainty	Uncertainty, unpredictability
Risk	Hazard, danger, threat, risk
Resilience *for what purpose?*	
Prevention	prevention, avoidance, circumvention
Adaptation	Adaptation, reorganization, transformation, adjustment, flexibility, innovation
Mitigation	Mitigation, consequence management
Improvement	Improvement, growth
Recovery	Recovery, reconstruction, bouncing back/ forward, repairment

The index proposed by Henry, D. [62] considers the proportion in resilience functionality $R_Q(t)$ between the functionality improvement from the damaged state and the reduction of functionality from the normal state. The restoration speed is not considered in this approach. This parameter is defined in Eq. (8.5).

$$R_Q(t) = \frac{Q(t) - Q(t_D)}{Q(t_E) - Q(t_D)} \qquad (8.5)$$

The concept to compare the actual performance $Q_{actual}(t)$ and the target performance $Q_{target}(t)$ as an instantaneous resilience index, R_{inst}, was advised by Ouyang, M. [63] as defined in Eq. (8.6).

$$R_{inst} = \frac{\int_0^T Q_{actual}(t)\,dt}{\int_0^T Q_{target}(t)\,d} \qquad (8.6)$$

where T is an observation period.

The strategies for resilience enhancement can be categorized, as is shown in Fig. 8.5, along two axes where strategy is either focused on the system's resilience or on the resilience of its governance, or on system structure or dynamics. The distinction between system structure and dynamics corresponds to a focus on the processes and organization that maintain resilience, with that of dynamic interacting systems. The distinction between strategies focused on the system – to be governed – and the governance system can also be interpreted as distinguishing between whether a system is being approached from a more analytical scientific perspective vs. a management or governance perspective [64]. As an example, the strategy for improving New Space resiliency by managing connectivity level is proposed by Pelton, J. N. [65]. However, the question "how to use analytical information in policy process" is addressed in the next section.

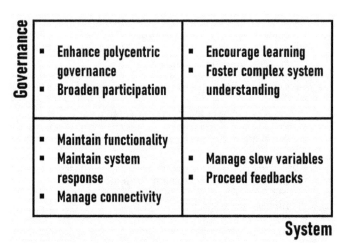

Figure 8.5: Strategies for enhancing resilience [66]

8.3   FROM SCIENCE TO RISK GOVERNANCE

The concept of *governance* emerged from Greek and Latin languages carrying connotations of the *art of steering* and *art of governing*. Depending on the nature of the system, a large spectrum of governance perspectives exist. The main ones are: process-centric, structure-centric, state-centric, hybrid, corporate governance, international security, and others. These perspectives determine three basic governance attributes:

- *direction* – which defines vision that supports consistent decisions, action, interpretation and strategic priorities;

- *accountability* – which involves ensuring efficient strategic resource utilization, performance monitoring, and exploration of aberrant conditions;

- *oversight* – which involves providing control, communication, coordination, and integration of systems and their components.

The speed at which the governance practices are established depends on contextual factors: technology characteristics including the number of elements and connectivity level [67], the total number of parts [68], the existence of appropriate measurement technique [69], the total amount of information [70]. The governance practices establishment is far from a straightforward linear process; it deals with the need to assess "unknown unknowns". Moreover, "unknown unknowns" can prevail over "known unknowns", such as in the case of space debris assessment.

Resilience building starts with defining a reasonable worst-case scenario. Stakeholders must be prepared to ensure critical infrastructure resilience at a reasonable cost. Early protective steps for space security are described by Pelton, J. N. [71]. They include use of an air traffic control system analogy and creation of an insurance scheme for space debris similar to launch insurance (more information on the risk transfer mechanism is given in Section 8.5).

The following recommendations for long-term resilience are proposed by authors. The first five recommendations are focused on disaster management, while others focus on improving a system's resiliency.

RECOMMENDATION I – The common understanding of a hazard and corresponding threats is shared using realistic scenarios. Normally, three return periods are studied: 1-in-10 years, 1-in-30 years and 1-in-100 years. It is known that risk is a function of the probability of occurrence and the estimated severity. The hazard's intensity increases as the return period increases. The level of risk (expressed in economic losses) is determined by the scenario probability. Therefore, the importance of using a consistent set of realistic scenarios is stressed. The examples of referenced scenarios are those published by Lloyd's of London on an annual basis in a realistic disaster scenario report. The 1-in-15 years is currently presented [72].

RECOMMENDATION II – It is suggested to consider events with recurrence intervals longer than 100 years as well. It was concluded that a wise emergency management practice to develop emergency operations plans based on the worst-case scenario according to historical data [73]. However, these data sets are currently missing for space debris hazard assessment, since the age of space activities is much shorter than 100 years![4]

RECOMMENDATION III – The results achieved within the previous step should be pushed further. Mitigation strategies based on a common scenario should be implemented across each policy area.

RECOMMENDATION IV – The need for developing, implementation and practising emergency operation plans before disaster strikes is emphasized. These emergency plans should describe emergency repair and recovery actions, assign responsibilities, identify resources and address coordination and communication [74]. The relevant plans should be prepared for both spaceborne and interconnected ground-based infrastructures. It was shown that the lack of direct interoperability with mutual aid resources and non-traditional responders is a notorious disaster response challenge [75].

RECOMMENDATION V – Space assets should be designed by considering prioritization of critical customers reliable operation. The example of how critical customers can be described, and how this information can be integrated in emergency management is given in Tab. 8.3.

Table 8.3: Example of how to prioritize critical customers and integrate this information in emergency planning

Customer				Space Infrastructure			
Designation	Type	Location	Outage Consequences	Space Asset	Consequences	Capabilities	Needs
-	-	-	-	-	-	-	-

RECOMMENDATION VI – The transition from hardening system assets and facilities to building resilience of a whole system is proposed. Traditional hazard mitigation strategies are focused on strengthening components based on the assessed risk level. These measures may be prohibitively expensive or impractical [76]. Building resiliency instead would not focus on preventing damage from a single hazard, but

---

[4]The Space Age is generally considered to have begun with Sputnik 1 in 1957.

rather to enable infrastructure to continue functioning when critical components are failed and swiftly return to operation after a disruption.

RECOMMENDATION VII – Consider the risk of increased space debris pollution in investment analyses not only for space asset design, but also for interconnected ground-based infrastructures.

One of the challenges of improving governance against space debris is effective international cooperation. Dealing with this topic requires collective action by the spacefaring nations and those who rely on reliable space system operation. The lack of national regulation and international norms for the responsible conduct of new kinds of space activities is proving to be a much larger challenge than the technical issues associated with developing such capabilities [77]. In this case, the governance problem should be studied as the system of systems comprised of nations with different attributes. With respect to infrastructure protection, the choice of subsystems can be as follows: economy, demographics, nature, politics, culture. Furthermore, each subsystem has subsections with a set of indicators. The method for assessing infrastructure vulnerability as a function of governance is given by Gheorghe, A. [78].

While critical infrastructure operators, owners and governments agree on the need for resilience building, the views on the levels of resilience may differ. The governing philosophies and policy documents vary significantly in resilience assessment. Alberts, D. [79] states the need to consider resilience across a broad spectrum of categories, including physical and information systems infrastructure as well as cognitive and social systems and frameworks.

It is particularity of interest to see how the United Nations Office for Outer Space Affairs (UNOOSA) applies the principles of resilience to the emerging threat of space debris for good governance of this issue in the coming years.

For studying this, we employed the use of the resilience matrix framework introduced by Linkov I. et al. [80] to compare temporal and spatial scales of resilience across UNOOSA's Committee on the Peaceful Uses of Outer Space (COPUOS), Scientific and Technical Subcommittee (STS), and Legal Subcommittee (LSC) from their published annual reports after the millennium based on the criteria that: the reports are publicly available and contain the keyword "Space Debris".

Space debris has been addressed in the documents[5] published by UNOOSA as early as in 1978, with almost a neglect in the 1980s and a reappearance in the 1990s, towards a boom after the millennium. The appearance of the keyword "Space Debris" in UNOOSA's (sub-)committees' annual reports after the millennium[6] is depicted in Fig. 8.6.

UNOOSA's annual reports containing the keyword "Space Debris" were then examined for the direct use of "Resilience" as a keyword in any form. The outcome is

---

[5]Publicly accessible on UNOOSA's Documents and Resolutions Database [81]

[6]In 2020, the 63rd session of Committee on the Peaceful Uses of Outer Space (COPUOS) and the 59th session of the Legal Subcommittee (LSC) did not take place on schedule due to the coronavirus outbreak.

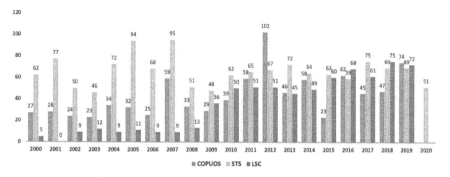

Figure 8.6: Appearance of the keyword "Space Debris" in UNOOSA (sub-) committees' annual reports after the millennium; the horizontal and vertical axes respectively represent the year and the number of appearances of the keyword

depicted in Fig. 8.7. As it is perceivable from this analysis, the resilience concept is an emerging topic. It is worth mentioning that there was no mention of the "Resilience" keyword in the treaties, conventions, and agreements made in the Old Space era including Outer Space Treaty (1967), Rescue Agreement (1968), Liability Convention (1972), Registration Convention (1975), Moon Agreement (1979). This shows that resilience was practically out of focus in the Old Space era.

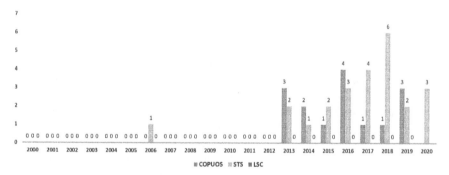

Figure 8.7: Appearance of the keyword "Resilience" in any form in UNOOSA (sub-)committees' annual reports after the millennium; the horizontal and vertical axes respectively represent the year and the number of appearances of the keyword

Following the method of Linkov I. et al [80], UNOOSA' publications on space debris along with the European Code of Conduct for Space Debris Mitigation (tagged EU), Inter-Agency Space Debris Coordination Committee (IADC)'s Space Debris Mitigation Guidelines, International Telecommunication Union (ITU)'s Recommendations, International Academy of Astronautics (IAA)'s Study on Space Traffic Management were scored for direct and indirect inclusion of temporal and spatial stages of resilience. The four temporal stages of resilience, i.e. "plan", "absorb", "recover" and "adapt" are defined as the following:

- *plan* – defines the steps taken by organizations to prepare critical functions and features of their operation for a universe of potential threats;

- *absorb* – comprises the capability of a system or organization to absorb the consequences of an acute shock or extended stress without breaking and maintaining a certain degree of function;

- *recover* – includes the time and resources needed for the system to recover its functionality post-shock;

- *adapt* – includes the capacity of an organization or system to "learn" and improve its capacity to absorb and recover from shocks based upon past experience.

Each publication was also scored based on the three primary spatial domains of resilience, including the "physical", "informational", and "social" aspects of resilience defined as the following:

- *physical* – showed that resilience was assessed within the context of physical infrastructure;

- *information* – revealed that resilience was discussed with regard to information flows and data moving up the system;

- *social* – showed that resilience was applied within the context of societal action and making society agile in the face of shock.

The outcome of this analysis is mapped in Fig. 8.8.

The resilience matrix shows that while all facets of resilience are considered across the collection of all (sub-)committees of UNOOSA and other international space debris related organizations, most focus is placed upon the "prepare" temporal stage on all three spatial domains, with a focus on "information". Likewise, the contents of the prepared guidelines does not consistently address the latter temporal important stages of resilience, "absorb", "recover" and "adapt". One notable exception is the work of COPUOS which covers all temporal and spatial domains. This shows that the United Nations Committee on the Peaceful Uses of Outer Space (COPUOS) is serving as a knowledge broker helping to share and push forward the strategic thinking of the resilience concept in the New Space era.

8.4   INTRODUCING SPACE DEBRIS AS A SYSTEMIC RISK

Often, space debris risk is defined as an emerging risk, since it is difficult to quantify and its potential impact on business is not sufficiently taken into account. This is based on the following promises: the rapid change of space debris environment, the growing inter-connectivity between space and ground-based infrastructures, the development of new technologies for problem assessing and mitigation, though "systemic risk"

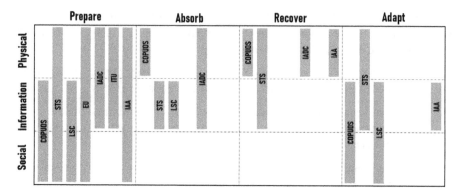

Figure 8.8: Resilience matrix showing the direct temporal (plan and prepare for, absorb, recover from, and adapt) and spatial (social, information, physical) domains of resilience for space debris related publications of UNOOSA (COPUOS, STS, LSC), European Code of Conduct for Space Debris Mitigation (tagged EU), Inter-Agency Space Debris Coordination Committee (IADC)'s Space Debris Mitigation Guidelines, International Telecommunication Union (ITU)'s Recommendations, International Academy of Astronautics (IAA)'s Study on Space Traffic Management

approach is a more appropriate concept for space debris risk assessment. Systemic risk is also sometimes called network risk, since it emerges from complex non-linear cause-effect interactions among individual elements or agents with different and often conflicting interests. In addition, each element or agent is characterized by its own risk portfolio. The examples of systemic risks are the financial crisis of 2008, pandemics, cybersecurity, global climate change. Systemic risks can trigger unexpected large-scale changes to a system or imply uncontrollable large-scale threats to it [82]. In other words, they tend to be *fat-tailed*.

The lack of knowledge about interdependencies requires advanced approaches to problem solving through risk thinking. The traditional linear methods have limited application. The concept of "femtorisks" stresses the importance of challenging standard approaches for risk assessment [83]. Because in a systems-approach there may be many competing solutions with no clear best, the challenge for their governance is to assure transparency, accountability, and inclusiveness of the risk management process, and effectiveness, stationarity, equity, and sustainability of the outcome [84]. One of the solutions is the principle of collective responsibility proposed by Helbing, D. [85].

Compared to critical risks, systemic risks do not attract the same attention and tend to be underestimated. The OECD defines critical risks as a rapid-onset event that pose the strategically significant risks as a result of their probability and likelihood. Another spaceborne hazard, space weather, was defined as a future global shock by the OECD [86]. Therefore, the concept of critical infrastructure resilience with respect to space threats is mainly focused on ground-based critical infrastructures, especially on power grids, which is the current backbone of modern critical infrastructures [87, 88]. Governments established the programs for assessing space weather risk, and this risk is included in the national risk portfolio of several countries (USA, Canada, Finland, Sweden, Norway, United Kingdom, Germany, the Netherlands, Hungary). They specified the tasks that will lead to improvements in policies, practices and procedures for decreasing vulnerability.

Recently, the OECD has developed a framework for good governance of critical infrastructure resilience that considers the system and interconnected risks to critical infrastructures [89]. This framework can be a good starting point for creating practices of space debris risk assessment. However, two points should be noted. The first point is stated by Bresch, D. N. [90]:

> [a] critical element of robust foresight includes innovation and experimentation.

The second point is that enhancing resilience to systemic risk is possible within systems thinking. Systems thinking is an inherent assumption within the complexity theory [91].

8.5   TOWARDS RISK TRANSFER. CONCLUSIONS

Dealing with the space debris risk has become a day-to-day business for space asset owners and operators. Moreover, there is an overarching societal interest in avoiding catastrophic impacts on the society well-being due to the collisions. Experience shows that the time after a disastrous event is the window of opportunities for implementing actions. The relevant stakeholders – in order to avoid the same loss in the future – are eager to implement actions with higher cost in order to boost a long-term resilience. However, the experience of catastrophic event mitigation (e.g. eruption of Eyjafjallajökull volcano in 2010, Japan tsunami 2011, hurricane Sandy in 2012, and others) proved that prevention is more beneficial than mitigation. Overall, the risk strategies can be classified with respect to the financial impact and likelihood of the event's occurrence as follows [92]:

- **Accept the risk** – activities responding to a practice of monitoring the risk if the cost-benefit analysis determines that the cost to mitigate the risk is higher than the cost to bear the risk. The best response in this case is to accept the risk.

- **Transfer the risk** – activities with low probability of occurring, but with a large financial impact. The best response is to transfer a portion or all of the risk to a third party by purchasing insurance, hedging, outsourcing, or entering into partnerships.

- **Mitigate the risk** – activities with a likelihood of occurring, but a small financial impact. The best response is to use management control systems to reduce the risk of potential loss.

- **Avoid the risk** – activities with a high likelihood of loss and large financial impact. The best response is to avoid the activity.

Risk transfer from policyholders to insurance, and later to reinsurance companies is an effective mechanism for ensuring financial stability. In other words, pooling risks reduces the uncertainty of expected loss over a given period of time. It helps to preserve the continuity of business especially in the presence of extensive events. Babylonian King Hammarubi's code (1750 B.C.) was the first record of insurance. Across the centuries, merchants were forming a pool to spread the loss. Enhancing maritime information exchange in Lloyd's Coffee House led to underwriting development. The first record is dated 1757. Large fires across Europe drove the development of the reinsurance sector: Cologne Re after the Hamburg Fire of 1842; and Swiss Reinsurance Company Ltd. (Swiss Re) after the Glarus Fire of 1861. These severe events demonstrated the need for reinsurance, although many reinsurance companies were in fact founded to prevent the outflow of reinsurance premiums from local economies to foreign ones [93].

Event scenario is the basis of the catastrophe models, which are used for specifying the amount of risk to be transferred and the price for it. A catastrophe model is a computerized system that generates a robust set of simulated events and estimates the magnitude, intensity and location of the event to evaluate the amount of damage and calculate the insured loss as a result of a catastrophic event [94]. The identification of key catastrophe models development milestones is given by Grossi, P. [95]. Catastrophe models are used by all three parties: insurance companies – who underwrite the original policy, reinsurance brokers – who works on behalf of the insurer to transfer risk to one or more reinsurances via reinsurance policies, and reinsurance companies. The graphical representation of a catastrophe model structure is given in Fig. 8.9.

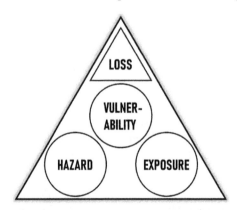

Figure 8.9: Catastrophe model structure

Any catastrophe model consists of four components:

(1) *Hazard* – is a process, phenomenon or human activity that may cause loss of life, injury or other health impacts, property damage, social and economic disruption or environmental degradation. Two types of hazard models are used:

deterministic (scenario models) – which are determined by assumed initial modelling conditions like historical hazard events or *what-if* scenarios, or probabilistic models – which estimate the probability of an event of a given severity. Contrary to deterministic models, probabilistic models can correlate spatial and temporal risks for a credibly scaled event. They combine historical data with theoretical and statistical models. The space debris peril is characterized by limited data points compared to other natural and technical hazards. However, even for more frequent hazards, insurance companies have only 10- or 20-years time-series of claims. It is due to the fact that the underlying trends – such as exposure landscape, infrastructure reliability standards, construction/maintenance costs – change.

(2) *Exposure* data – is the primary input to catastrophe model. The exposure can be evaluated using a combination of geospatial mapping and probabilistic modelling. The quality of exposure data varies greatly in various parts of the world and the industry lines. Along with object location, sum insured, primary and secondary modifiers should be considered. The S-curve approach is normally used for dealing with the change of the exposure amount through the life-cycle. Catastrophe models can assess the economic impact by associating an event-based model with an economic exposure database [96].

(3) *Vulnerability* component – links hazard and exposure components. Most vulnerability models are arranged as a series of damage functions, which enable look-up between hazard intensity and estimated damage as a ratio of total value [97]. Damage ratio $DR(i, j)$ for a i object from a j peril can be represented as shown in Eq. (8.7).

$$DR(i, j) = \frac{Repair\ Cost\ (i, j)}{Total\ Repair\ Cost\ (i, j)} \tag{8.7}$$

(4) *Loss* module – is an output of a vulnerability module. It translates infrastructure damage into costs covered by insurance. It is usually expressed as ground up loss which is the entire amount of an insurance loss, including deductibles, before application of any retention or reinsurance; retained (client) loss, which is the loss to be insured; gross loss, which is the amount of a ceding company's loss irrespective of any reinsurance recoveries due. It is essential that the loss reflects the impact of insurance products and mechanisms. The main functions of a loss module are: reflecting any insurance and reinsurance policy conditions; aggregating the location-coverage losses to higher levels (e.g. policy or country levels); back allocating the impact of higher-level policy structure to lower levels so that impact of higher-level structures can be understood and summarized at a more detailed level; calculating summary metrics such as Average Annual Loss (AAL), Occurrence Exceedance Probability (OEP) and Aggregate Exceedance Probability (AEP) curves [97]. Afterwards, the catastrophe model should pass the iteration validation process. At the end, there should be a clear understanding of:

- Which spatial and temporal resolutions were achieved? Is this high enough?
- How are footprints calibrated?
- Which historical events are considered? For model creation? For its validation?
- Which hazard metrics are used?
- How does loss data vary among territories and industries?
- What is the exposure data quality?
- What are the sources and the range of uncertainties?

Historically, insurers provide a proven and tested insurance product for property insurance against the failure of that satellite during launch or operation and will typically recoup only the cost of the satellite, not the loss of future revenue [98]. The evolving market and environment conditions dictate the need of new insurance products. One of the leaders in the space insurance industry, Swiss Re, states that the main challenge is to find a way to reconcile the carefully crafted insurance product that responds to the bespoke requirements of the constellation operators with the heightened risk that the deployment of such a large number of new satellites clustered in an already densely populated LEO increasingly poses [99]. Orbital space is considered to be very large, and the relative size of space assets is in contrast very small. Besides the fast increase of the number of satellites, the collision possibility is still relatively low, though the severity of the consequences is high. The tools for assessing space debris risk are getting continuously developed including the improvement of the Meteoroid and Space debris Terrestrial Environment Reference (MASTER) [100]. According to the risk strategies descriptions given above, one of the best solutions is a risk transfer. However, the insurance is not helpful until relevant stakeholders clearly answer: "what are we trying to avoid?" and "what problem are we trying to solve?".

The space debris problem is inseparable from the problem of improving critical infrastructure resilience. The on-going dialogues among industry, scientists, policy makers and economists should give more information on: "how extreme can future events be?", "what is the expected frequency of such events?", and "what damage can be expected?". Realistic and adequate answers to these questions are the silver bullets for defining appropriate resilience enhancement strategy including governance options. It should be noted that any steps taken to reduce long-term risks also minimize the potential increase of short-term risks.

ACRONYM

AAL Average Annual Loss

AEP Aggregate Exceedance Probability

COMM Communication Satellite

COPUOS Committee on the Peaceful Use of Outer Space

ESA European Space Agency

EU European Union

GDP Gross Domestic Product

GEO Geostationary Earth Orbit

GMD Geomagnetic disturbance

GNSS Global Navigation Satellite System

GVA Gross Value-Added

IAA International Academy of Astronautics

IAC International Astronautical Congress

IADC Inter-Agency Space Debris Coordination Committee

ITU International Telecommunication Union

LEO Low Earth Orbit Region

LSC Legal Subcommittee of COPUOS

MASTER Meteoroid and Space Debris Terrestrial Environment Reference

MEO Medium Earth Orbit Region

NACE Nomenclature des Activités Économiques dans la Communauté Européenne

NASA National Aeronautics and Space Administration

NEO Near-Earth Object

NIPP National Infrastructure Protection Plan

NOAA National Oceanic and Atmospheric Administration

OECD Organization for Economic Cooperation and Development

OEP Occurrence Exceedance Probability

STS Scientific and Technical Subcommittee of COPUOS

UNDRR United Nations Office for Disaster Risk Reduction

UNOOSA United Nations Office for Outer Space Affairs

WIOD World Input-Output Database

GLOSSARY

Aggregate Exceedance Probability (AEP): The probability of the sum of event losses in a year exceeding a certain level.

Average Annual Loss (AAL): The expected loss cost over a one-year time period.

Cost/risk criteria: A criteria of the loss and required cost for system reinforcement.

Critical infrastructure: An infrastructure that provides an essential support for economic and social well-being, public safety and the functioning key government responsibilities, such that disruption or destruction of the infrastructure would result in catastrophic and far-reaching damage.

Critical risk: A rapid-onset event that pose the strategically significant risk as a result of its probability and likelihood.

Damage ratio: The estimated repair cost of an asset at risk divided by the replacement cost of an asset.

Emerging risk: An issue that is perceived to be potentially significant but which may not be fully understood.

Exposure data: The data representing the assets to be modelled.

Governance: The processes, controls and oversight put in place for ensuring that catastrophe risk is properly managed.

Hazard: A process, phenomenon or human activity that may cause loss of life, injury or other health impacts, property damage, social and economic disruption or environmental degradation. Geomagnetic disturbance (GMD) is considered as a hazard and a blackout caused by the GMD is a disaster.

Liability Convention: Convention on International Liability for Damage Caused by Space Objects (1972).

Mitigation: Actions taken to reduce the impact of a hazard.

Moon Agreement: Agreement Governing the Activities of States on the Moon and Other Celestial Bodies (1979).

Realistic disaster scenario: Catastrophe scenario used for exposure management.

Resilience: The ability of households, communities and nations to absorb and recover from shocks, whilst positively adapting and transforming their structures and means for living in the face of long-term stresses, change and uncertainty.

Risk: An expected loss due to a particular hazard for a given area and a reference period.

Registration Convention: Convention on Registration of Objects Launched into Outer Space (1975).

Rescue Agreement: Agreement on the Rescue of Astronauts, the Return of Astronauts and the Return of Objects Launched into Outer Space (1968).

Scenario: A representation of a possible event based on scientific analysis or expert knowledge.

Systemic risk: A risk that emerges from complex interactions among individual elements or agents, which are attributed with their own risk portfolio.

FURTHER READING

Johnson-Freese, J. (2007). *Space as a Strategic Asset.* Columbia University Press.

Burch, R. (2019). *Resilient Space Systems Design: An Introduction.* CRC Press.

Linkov, I., & Trump, B. D. (2019). *Resilience and Governance.* In The Science and Practice of Resilience (pp. 59–79). Springer, Cham.

Lowe, C. J., & Macdonald, M. (2020). *Space mission resilience with inter-satellite networking.* Reliability Engineering & System Safety, 193, 106608.

Bibliography

[1] A. A. Ghorbani and E. Bagheri. The State of the Art in Critical Infrastructure Protection: a Framework for Convergence. *International Journal of Critical Infrastructures*, 4(3):215–244, 2008.

[2] W. J. Clinton. Executive order 13010 – Critical Infrastructure Protection. *Federal Register*, 61(138):37347–37350, 1996.

[3] US Congress. Uniting and Strengthening America by Providing Appropriate Tools Required to Intercept and Obstruct Terrorism (USA PATRIOT ACT) Act of 2001. *Public Law*, pages 107–56, 2001.

[4] K. Gordon and M. Dion. Protection of "Critical Infrastructure" and the Role of Investment Policies Relating to National Security. *Investment Division, Directorate for Financial and Enterprise Affairs, Organisation for Economic Cooperation and Development, Paris*, 75116, 2008.

[5] O. Sokolova and V. Popov. Critical Infrastructure Exposure to Severe Solar Storms. *Safety and Reliability: Methodology and Applications*, pages 1327–1340, 2017.

[6] C. Directive. 114/EC of 8 December 2008 on the Identification and Designation of European Critical Infrastructures and the Assessment of the Need to Improve their Protection. *Official Journal of the European Union L*, 345(75):23–12, 2008.

[7] E. Zio. Challenges in the Vulnerability and Risk Analysis of Critical Infrastructures. *Reliability Engineering & System Safety*, 152:137–150, 2016.

[8] I. Granic and A. V. Lamey. The Self-Organization of the Internet and Changing Modes of Thought. *New Ideas in Psychology*, 18(1):93–107, 2000.

[9] S. Mittal. Emergence in Stigmergic and Complex Adaptive Systems: A Formal Discrete Event Systems Perspective. *Cognitive Systems Research*, 21:22–39, 2013.

[10] T. De Wolf and T. Holvoet. Emergence Versus Self-Organisation: Different Concepts but Promising When Combined. In *International Workshop on Engineering Self-Organising Applications*, pages 1–15. Springer, 2004.

[11] S. M. Rinaldi, J. P. Peerenboom, and T. K. Kelly. Identifying, Understanding, and Analyzing Critical Infrastructure Interdependencies. *IEEE Control Systems Magazine*, 21(6):11–25, 2001.

[12] E. Kilpua, H. E. Koskinen, and T. I. Pulkkinen. Coronal Mass Ejections and Their Sheath Regions in Interplanetary Space. *Living Reviews in Solar Physics*, 14(1):5, 2017.

[13] R. B. Horne, M. W. Phillips, S. A. Glauert, N. P. Meredith, A. Hands, K. A. Ryden, and W. Li. Realistic Worst Case for a Severe Space Weather Event Driven by a Fast Solar Wind Stream. *Space Weather*, 16(9):1202–1215, 2018.

[14] M. J. Egan. Anticipating Future Vulnerability: Defining Characteristics of Increasingly Critical Infrastructure-like Systems. *Journal of Contingencies and Crisis Management*, 15(1):4–17, 2007.

[15] John S. Foster J., E. Gjelde, W. R. Graham, R. J. Hermann, H. M. Kluepfel, R. L. Lawson, G. K. Soper, L. L. Wood, and J. B. Woodard. Report of the Commission to Assess the Threat to the United States from Electromagnetic Pulse (EMP) Attack: Critical National Infrastructures. Technical report, ELECTRO-MAGNETIC PULSE (EMP) COMMISSION MCLEAN VA, 2008.

[16] I. Eusgeld, D. Henzi, and W. Kröger. Comparative Evaluation of Modeling and Simulation Techniques for Interdependent Critical Infrastructures. *Scientific Report, Laboratory for Safety Analysis, ETH Zurich*, pages 6–8, 2008.

[17] R. K. Garrett Jr., S. Anderson, N. T. Baron, and J. D. Moreland Jr. Managing the Interstitials, a System of Systems Framework Suited for the Ballistic Missile Defense System. *Systems Engineering*, 14(1):87–109, 2011.

[18] H. A. Rahman, J. R. Marti, and K. D. Srivastava. Quantitative Estimates of Critical Infrastructures' Interdependencies on the Communication and Information Technology Infrastructure. *International Journal of Critical Infrastructures*, 7(3):220–242, 2011.

[19] E. Koks, R. Pant, S. Thacker, and J. W. Hall. Understanding Business Disruption and Economic Losses Due to Electricity Failures and Flooding. *International Journal of Disaster Risk Science*, pages 1–18, 2019.

[20] R. A. Kerr. *Signs of Success in Forecasting El Niño*. American Association for the Advancement of Science, 2002.

[21] H. R. Hertzfeld, R. A. Williamson, S. Dick, and R. Launius. The Social and Economic Impact of Earth Observing Satellites. *Societal Impact of Space Flight*, pages 237–264, 2007.

[22] C. Jolly and G. Razi. *The Space Economy at a Glance*. OECD Publishing, 2007.

[23] S. Mukherjee, R. Nateghi, and M. Hastak. A Multi-Hazard Approach to Assess Severe Weather-Induced Major Power Outage Risks in the US. *Reliability Engineering & System Safety*, 175:283–305, 2018.

[24] World Economic Forum. *The Global Risks Report 2018, 13th Edition*. 2018, Geneva.

[25] Economics London. The Economic Impact on the UK of a Disruption to GNSS. *Showcase Final Report, UK Space Agency*, 2017.

[26] G. Bachner, B. Bednar-Friedl, S. Nabernegg, and K. W. Steininger. Economic Evaluation Framework and Macroeconomic Modelling. In *Economic Evaluation of Climate Change Impacts*, pages 101–120. Springer, 2015.

[27] E. Oughton, J. Copic, A. Skelton, V. Kesaite, Z. Yeo, S. Ruffle, and D. Ralph. Helios Solar Storm Scenario. *Cambridge Risk Framework Series. Cambridge, UK: Centre for Risk Studies, University of Cambridge.*, 2016.

[28] P. B. Dixon and B. R. Parmenter. Computable General Equilibrium Modelling for Policy Analysis and Forecasting. *Handbook of Computational Economics*, 1:3–85, 1996.

[29] K. Akao and H. Sakamoto. A Theory of Disasters and Long-Run Growth. *Journal of Economic Dynamics and Control*, 95:89–109, 2018.

[30] S. Kelly. Estimating Economic Loss from Cascading Infrastructure Failure: A Perspective on Modelling Interdependency. *Infrastructure Complexity*, 2(1):7, 2015.

[31] M. Timmer, B. Los, R. Stehrer, and G. de Vries. An Anatomy of the Global Trade Slowdown Based on the WIOD 2016 Release. Technical report, Groningen Growth and Development Centre, University of Groningen, 2016.

[32] M. E. Kahn. The Death Toll from Natural Disasters: The Role of Income, Geography, and Institutions. *Review of Economics and Statistics*, 87(2):271–284, 2005.

[33] H. Toya and M. Skidmore. Economic Development and the Impacts of Natural Disasters. *Economics Letters*, 94(1):20–25, 2007.

[34] G. Felbermayr and J. Gröschl. Naturally Negative: The Growth Effects of Natural Disasters. *Journal of Development Economics*, 111:92–106, 2014.

[35] E. Cavallo and I. Noy. Natural Disasters and the Economy – A Survey. *International Review of Environmental and Resource Economics*, 5(1):63–102, 2011.

[36] E. Skoufias. Economic Crises and Natural Disasters: Coping Strategies and Policy Implications. *World Development*, 31(7):1087–1102, 2003.

[37] C. J. Schrijver. Socio-Economic Hazards and Impacts of Space Weather: The Important Range Between Mild and Extreme. *Space Weather*, 13(9):524–528, 2015.

[38] C. J. Schrijver, K. Kauristie, A. D. Aylward, C. M. Denardini, S. E. Gibson, A. Glover, N. Gopalswamy, M. Grande, M. Hapgood, and D. Heynderickx. Understanding Space Weather to Shield Society: A Global Road Map for 2015–2025 Commissioned by COSPAR and ILWS. *Advances in Space Research*, 55(12):2745–2807, 2015.

[39] O. Sokolova and M. Madi. A View of the New Space Sector Resilience. *Proceedings of the 30th European Safety and Reliability Conference, ESREL 2020*.

[40] G. Purdy. ISO 31000: 2009–Setting a New Standard for Risk Management. *Risk Analysis: An International Journal*, 30(6):881–886, 2010.

[41] B. S. Blanchard, W. J. Fabrycky, and W. J. Fabrycky. *Systems Engineering and Analysis*, volume 4. Prentice Hall Englewood Cliffs, NJ, 1990.

[42] L. Skyttner. *General Systems Theory: Problems, Perspectives, Practice*. World Scientific, 2005.

[43] P. F. Katina, C. B. Keating, and A. V. Gheorghe. Cyber-Physical Systems: Complex System Governance as an Integrating Construct. In *Proceedings of the 2016 Industrial and Systems Engineering Research Conference. Anaheim, CA: IISE*, 2016.

[44] O. Renn. White Paper on Risk Governance: Toward an Integrative Framework. In *Global Risk Governance*, pages 3–73. Springer, 2008.

[45] T. Aven. The Call for a Shift from Risk to Resilience: What Does It Mean? *Risk Analysis*, 39(6):1196–1203, 2019.

[46] B. Ramalingam, H. Jones, T. Reba, and J. Young. *Exploring the Science of Complexity: Ideas and Implications for Development and Humanitarian Efforts*, volume 285. Overseas Development Institute London, 2008.

[47] R. Klein, R. Nicholls, and F. Thomalla. Resilience to Natural Hazards: How Useful is This Concept? *Global Environmental Change Part B: Environmental Hazards*, 5(1):35–45, 2003.

[48] C. S. Holling. Resilience and Stability of Ecological Systems. *Annual Review of Ecology and Systematics*, 4(1):1–23, 1973.

[49] R. Francis and B. Bekera. A Metric and Frameworks for Resilience Analysis of Engineered and Infrastructure Systems. *Reliability Engineering & System Safety*, 121:90–103, 2014.

[50] United Nations Office for Disaster Risk Reduction (UNISDR). *UNISDR Terminology on Disaster Risk Reduction*. Geneva: United Nations, 2009.

[51] Cabinet Office. *Keeping the Country Running: Natural Hazards and Infrastructure.* London, 2011.

[52] M. Panteli and P. Mancarella. The Grid: Stronger, Bigger, Smarter?: Presenting a Conceptual Framework of Power System Resilience. *IEEE Power and Energy Magazine*, 13(3):58–66, 2015.

[53] M. Panteli, D. N. Trakas, P. Mancarella, and N. D. Hatziargyriou. Power Systems Resilience Assessment: Hardening and Smart Operational Enhancement Strategies. *Proceedings of the IEEE*, 105(7):1202–1213, 2017.

[54] D. J. Snowden and M. E. Boone. A Leader's Framework for Decision Making. *Harvard Business Review*, 85(11):68, 2007.

[55] World Economic Forum. Building Resilience to Natural Disasters: A Framework for Private Sector Engagement. 2008.

[56] B. Walker and D. Salt. *Resilience Practice: Building Capacity to Absorb Disturbance and Maintain Function.* Island Press, 2012.

[57] E. Hollnagel. *Safety–I and Safety–II: The Past and Future of Safety Management.* Ashgate Publishing, Ltd., 2014.

[58] J. D. Moteff. *Critical Infrastructures: Background, Policy, and Implementation.* DIANE Publishing, 2010.

[59] United Nations & World Bank. *Natural Hazards, Unnatural Disasters: The Economics of Effective Prevention.* World Bank Publication, Washington, DC, 2010.

[60] P. E. Roege, Z. A. Collier, J. Mancillas, J. A. McDonagh, and I. Linkov. Metrics for Energy Resilience. *Energy Policy*, 72:249–256, 2014.

[61] S. Hosseini, K. Barker, and J. E. Ramirez-Marquez. A Review of Definitions and Measures of System Resilience. *Reliability Engineering & System Safety*, 145:47–61, 2016.

[62] D. Henry and J. E. Ramirez-Marquez. Generic Metrics and Quantitative Approaches for System Resilience as a Function of Time. *Reliability Engineering & System Safety*, 99:114–122, 2012.

[63] M. Ouyang and L. Duenas-Osorio. Multi-dimensional Hurricane Resilience Assessment of Electric Power Systems. *Structural Safety*, 48:15–24, 2014.

[64] A. E. Quinlan, M. Berbés-Blázquez, L. J. Haider, and G. D. Peterson. Measuring and Assessing Resilience: Broadening Understanding Through Multiple Disciplinary Perspectives. *Journal of Applied Ecology*, 53(3):677–687, 2016.

[65] J. N. Pelton. Resiliency, Reliability, and Sparing Approaches to Small Satellite Projects. *Handbook of Small Satellites: Technology, Design, Manufacture, Applications, Economics and Regulation*, pages 1–15, 2019.

[66] R. Biggs, M. Schlüter, and M. L. Schoon. *Principles for Building Resilience: Sustaining Ecosystem Services in Social-Ecological Systems.* Cambridge University Press, 2015.

[67] J. T. Macher. Technological Development and the Boundaries of the Firm: A Knowledge-based Examination in Semiconductor Manufacturing. *Management Science*, 52(6):826–843, 2006.

[68] K. Singh. The Impact of Technological Complexity and Interfirm Cooperation on Business Survival. *Academy of Management Journal*, 40(2):339–367, 1997.

[69] J. S. Brown and P. Duguid. Knowledge and Organization: A Social-Practice Perspective. *Organization Science*, 12(2):198–213, 2001.

[70] E. Von Hippel. "Sticky Information" and the Locus of Problem Solving: Implications for Innovation. *Management Science*, 40(4):429–439, 1994.

[71] J. N. Pelton. A Path Forward to Better Space Security: Finding New Solutions to Space Debris, Space Situational Awareness and Space Traffic Management. *Journal of Space Safety Engineering*, 6(2):92–100, 2019.

[72] Lloyds. *RDS 2020 Realistic Disaster Scenarios Scenario Specification January 2020.* Lloyds, 2007.

[73] R. W. Perry and M. Lindell. *Emergency Planning.* Wiley, 2007.

[74] Direction Génerale de la Sécurité Civile et de la Gestion des Crises (DGSCGC). *Guide ORSEC Départament et Zonal: Mode d'Action Rétablissement et Approvisionnement d'Urgence des Réseaux Électricité, Communication Électroniques, Eau, Gaz Hydrocarbures.* Paris, 2015.

[75] US Fire Administration. *Operational Lessons Learned in Disaster Response.* Federal Emergency Management Agency, 2015.

[76] N. C. Abi-Samra. One Year Later: Superstorm Sandy Underscores Need for a Resilient Grid. *IEEE Spectrum*, 4:321–354, 2013.

[77] P. Martinez. Challenges for Ensuring the Security, Safety and Sustainability of Outer Space Activities. *The Journal of Space Safety Engineering*, 6:65–68, 2019.

[78] A. V. Gheorghe, D. V. Vamanu, P. F. Katina, and R. Pulfer. System of Systems Governance. In *Critical Infrastructures, Key Resources, Key Assets*, pages 93–130. Springer, 2018.

[79] D. S. Alberts and R. E. Hayes. Power to the edge: Command... control... in the information age. Technical report, Office of the Assistant Secretary of Defense, Command and Control Research Program (CCRP), 2003.

[80] I. Linkov, B. D. Trump, K. Poinsatte-Jones, P. Love, W. Hynes, and G. Ramos. Resilience at OECD: Current State and Future Directions. *IEEE Engineering Management Review*, 46(4):128–135, 2018.

[81] UNOOSA. Documents and Resolutions Database (accessed on May 25, 2020). https://www.unoosa.org/oosa/documents-and-resolutions/search.jspx?view=documents.

[82] D. Helbing. *Systemic Risks in Society and Economics. International Risk Governance Council.* 2010.

[83] A. B. Frank, M. Collins, S. Levin, A. Lo, J. Ramo, U. Dieckmann, V. Kremenyuk, A. Kryazhimskiy, J. Linnerooth-Bayer, and B. Ramalingam. Dealing with Femtorisks in International Relations. *Proceedings of the National Academy of Sciences*, 111(49):17356–17362, 2014.

[84] S. Jacobzone, C. Baubion, J. Radisch, S. Hochrainer-Stigler, J. Linnerooth-Bayer, W. Liu, E. Rovenskaya, and U. Dieckmann. Strategies to Govern Systemic Risks. OECD, 2020.

[85] D. Helbing. Globally Networked Risks and How to Respond. *Nature*, 497(7447):51–59, 2013.

[86] Organisation for Economic Co operation and Development (OECD). *OECD Reviews of Risk Management Policies – Future Global Shocks: Improving Risk Governance.* OECD Paris, France, 2011.

[87] O. Sokolova and V. Popov. Concept of Power Grid Resiliency to Severe Space Weather. In *International Conference on Intelligent Systems in Production Engineering and Maintenance*, pages 107–117. Springer, 2018.

[88] O. Sokolova, P. Burgherr, Ya. Sakharov, and N. Korovkin. Algorithm for Analysis of Power Grid Vulnerability to Geomagnetic Disturbances. *Space Weather*, 16(10):1570–1582, 2018.

[89] Organisation for Economic Co operation and Development (OECD). *Good Governance for Critical Infrastructure Resilience.* OECD Paris, France, 2019.

[90] D. N. Bresch. Shaping Climate Resilient Development: Economics of Climate Adaptation. In *Climate Change Adaptation Strategies – An Upstream-Downstream Perspective*, pages 241–254. Springer, 2016.

[91] E. Ostrom and M. A. Janssen. Multi-Level Governance and Resilience of Social-Ecological Systems. In *Globalisation, Poverty and Conflict*, pages 239–259. Springer, 2004.

[92] N. T. Sheehan. A Risk-based Approach to Strategy Execution. *Journal of Business Strategy*, 2010.

[93] P. Borscheid, D. Gugerli, and T. Straumann. *The Value of Risk: Swiss Re and the History of Reinsurance.* OUP Oxford, 2013.

[94] Lloyd's Market Association (LMA). *Catastrophe Modelling: Guidance for Non-Catastrophe Modelers.* 2013, London.

[95] P. Grossi. *Catastrophe Modeling: A New Approach to Managing Risk*, volume 25. Springer Science & Business Media, 2005.

[96] R. Gunasekera, O. Ishizawa, C. Aubrecht, B. Blankespoor, S. Murray, A. Pomonis, and J. Daniell. Developing an Adaptive Global Exposure Model to Support the Generation of Country Disaster Risk Profiles. *Earth-Science Reviews*, 150:594–608, 2015.

[97] K. Mitchell-Wallace, M. Jones, J. Hillier, and M. Foote. *Natural Catastrophe Risk Management and Modelling: A Practitioner's Guide*. John Wiley & Sons, 2017.

[98] V. A. Samson, J. D. Wolny, and I. Christensen. Can the Space Insurance Industry Help Incentivize the Responsible Use of Space? *Proceedings of 69th International Astronautical Congress (IAC), Bremen, Germany, 1–5 October 2018.*, 2018.

[99] P. Chrystal, D. Mcknight, and P. Meredith. New Space, New Dimensions, New Challenges: How Satellite Constellations Impact Space Risk. *Publication no. 1507500_18_EN, Swiss Re, Zürich, Switzerland*, 2018.

[100] V. Braun. Impact of Debris Model Updates on Risk Assessments. In *Proceedings of the 1st NEO and Space Debris Detection Conference*, 2019.

VII

Conclusion

CHAPTER 9

Pathways to Opportunities

Matteo Madi

S PACE is not unlimited, which is contrary to what we used to believe; but it is
[still] true that space offers enormous windows of opportunities. The substantial
shift from *Old Space* to *New Space* starts a dialogue on how resilient the space sector
is. According to the United Nations Office for Outer Space Affairs (UNOOSA), 70%
of space activities are now led by the private sector. In other words, 70% of space
activities are run under stiff marginal productivity constraints in a highly risky environment due to the fact that the current space situation is not properly managed,
and substantive ongoing collision risks exist in both Low Earth Orbit (LEO) and
Geostationary Earth Orbit (GEO). This is a sobering fact that shows the importance
of timely, accurate, comprehensive, highly-available and standards-based space situational awareness. While we are globally challenged by this peril today, extensive
leveraging of advanced algorithms, research, and crowd-sourcing and data fusion of
spacecraft operator and government and commercial data can provide us with the
critical capabilities and data required to address these space situational challenges.
Achieving long-term sustainability supports explosive growth in both the number of
active spacecrafts and our knowledge of the debris population within the decade.

As with the governance of other global sustainability issues, like climate change,
various hurdles stand in the way of finding sufficient and satisfying answers to make
our future sustainable. Successfully tackling space debris as a bidirectional sustainability risk is thus to acknowledge two key aspects: (1) it is difficult to accomplish,
requires a thoroughly collaborative international approach, and has to be recognized
as a sociotechnical challenge of global magnitude that societies might not [yet] be
fully equipped to engage with; (2) it is possible to identify this risk as a chance and
incentive to consider the role of outer space as an environment in the public perception
of and interaction with outer space and the space sector. However, space debris holds
one decisive advantage to be effectively contained as compared to other challenges of

environmental sustainability: its "remote nature" and the "unresolved societal status of outer space as an environment". This might give considerable leverage to adopt extended stakeholder participation in the process of space debris mitigation and removal, but also the future of New Space as a whole.

With renewed interest in spaceflight and satellite assets currently on the rise, public participation in finding solutions to the space debris problem can, and should be, fostered. On the side of space debris mitigation, this could, for example, mean intensified involvement of new groups of actors into evaluating downstream demand for New Space applications, prioritization of services to be delivered via orbits, or even the mission design of satellite constellations themselves. Similarly, regular public participation provides ample opportunity to address questions such as: "what levels of risk of in-orbit collisions are acceptable in relation to the value of the outer space environment?", "what are reasonable and justifiable thresholds for injury and damage by space debris re-entry?", "which missions and applications are most valuable for societies to be provided from orbit?". While these questions require strong expertise and professional experience to answer and put into policy, the stakeholder involvement could bring not only an "outside" view that potentially merits previously unconsidered aspects, but it also might strengthen the legitimacy of standards and policies of truly global reach.

Analysis shows that the conventional framework provides limited help when it comes to space debris. None of the five United Nations' treaties, conventions, and agreements made in the Old Space era including Outer Space Treaty (1967), Rescue Agreement (1968), Liability Convention, (1972), Registration Convention (1975), Moon Agreement (1979), address the issue of the creation and multiplication of space debris in a satisfactory manner. As a result several strategic documents have been developed, though legal uncertainty and controversy are here to stay. The Space Debris Mitigation Guidelines of the Inter-agency Space Debris Coordination Committee (IADC) and the United Nations' Committee for the Peaceful Use of Outer Space (COPUOS) contain essential elements of mitigation and benefit from a high level of implementation.

The concept of critical infrastructure resilience with respect to space threats is mainly focused on ground-based critical infrastructures. However, the New Space infrastructure has received attention as an emerging system, and is evolving as a new backbone system, the reliable operation of which is decisive for the society's well-being and economic stability. In terms of legal support, the United Nations' COPUOS serves as a knowledge broker helping to share and push forward the strategic thinking or resilience concept in the new Space era. Dealing with the space debris risk is gradually becoming a day-to-day business not only for space asset owners and operators. There is an overarching societal interest in avoiding catastrophic impacts due to the collisions. The space asset loss may overall have global impacts, and the impact can go beyond the borders of a certain country who owns the satellite.

Remediation is currently limited by the following legal difficulties: (1) the definition of what constitutes space debris, (2) what mechanisms would allow the removal of the debris without violating States' jurisdiction, (3) the liability which may arise in case of damage caused during remediation operations, and (4) sensitive questions relating to intellectual property and national security.

Overall, risks related to the space debris problem can be divided into two big categories: risk associated with space debris collision and risks related to space debris removal projects. The space debris risk is partially transferred to the insurance market. However, more adapted solutions could be considered in order to offer a better and greater protection in respect to debris mitigation or removal projects. A better legal environment will bring more comfort to insurance companies.

To solve the space debris problem and maintain sustainable and safe Earth orbits, it is important to reduce the future production of space debris by actively removing the existing, in-situ debris, given that it can lead to additional collisions and break-ups. Various technologies, such as in-situ identification of space debris, rendezvous and homing, and robotics, need to be improved to actively remove space debris. Simultaneously, we need to maintain the costs incurred in the implementation of these technologies at low levels, although they require increased autonomy and performance. A few ideas to harmonize the technological requirements and alleviate the pressure for their realization were introduced in this book. Nevertheless, it is believed that these may not be sufficient because the required technologies are widespreading. Therefore, researchers and engineers, particularly the younger generation, should be encouraged to tackle technologies for active debris removal. Sustainable and safe Earth orbits partly depend on noble and unique ideas that aim to mitigate these technical issues.

The topic of large-scale space debris is undergoing very rapid and vigorous development, especially taking into account the consequences of the collision of two satellites in 2009 and the anti-satellite weapons tests in near space. All large objects are catalogued, and their orbits are constantly monitored. In particular, the directions related to the development of flyby methods of such objects and the methods of their capture and fixation are actively studied. Approximately 11% of the currently catalogued objects are fragments formed during a spacecraft's lifecycle. The amount of debris generated during space hardware operation should be minimized. At present, space agencies undertake measures to prevent the emergence of such objects. Current international agreements for LEO and GEO stipulate that large space debris objects should be de-/re-orbited to disposal orbits at the end of their service life. However, such rules cannot be applied to objects that have become space debris prior to the adoption of these recommendations.

According to Swiss Re, the population of man-made objects in GEO has increased dramatically by roughly 40% in the course of the past eight years. While this increase is partly due to the launch of new satellites, that are still operational, it is mainly caused by the growth in the space debris population. The growth in the debris population is primarily due to satellite breakups. There has been an increase in almost every category of debris, as there is no natural cleansing mechanism (such as atmospheric drag) in GEO. On the positive side, compliance with re-orbiting guidelines has been improving in GEO and satellite servicing/removal capabilities appear to be emerging over the next decade.

The modern satellites deployed in LEO are either equipped with end-of-life disposal facilities (such as micro-electric propulsion systems (micro-EPS), reserved chemical fuel, etc.), or are planned to be de-orbited gradually due to atmospheric drag forces (valid for satellites in lower orbits). The constellations are planned to occupy higher LEO orbits, therefore any out-of-order satellite is potentially a source of debris if not equipped with sufficient manoeuvring propellant. Consequently, satellite constellations players should either equip their satellites with an end-of-life disposal kit or to find other cooperative solutions for safely de-orbiting their satellites.

From a business perspective, venture capitals look at a quick return of their investments, high profits and short times for space-based products and services to reach the markets. Disruptive technology innovation will be the only way for New Space companies to create new markets and remain competitive. Thereby, among companies who are offering On-Orbit Servicing (SSO) or Active Debris Removal (ADR) services, there are those who possess a more innovative technology with an acceptable TRL (Technology Readiness Level) which could be implemented in short-term for dysfunctional satellites, e.g. in GEO, as a niche immediate-market ready to be explored, and/or in a longer term to serve constellations in LEO. The crucial question is "what is the role of space debris removal players concerning the satellite constellations business?". In other words, "are satellite constellations players ready to partner with space debris removal players?" In only that case, a short-time benefit can be imagined. We should not forget that collisions in LEO may be catastrophic while they can be mission-terminating in GEO. That's another clue why space debris removal technologies designed for GEO operation are of high interest for the emerging space market. It should be noted that

> [the] cost of developing and maintaining infrastructures is justified by the mitigation of the impact resulting from the disruption of the daily economic activity.

It is worth mentioning that no space debris removal service is in operation today, though such service is of high interest for the large objects from highly populated orbits by different means. It is yet too early to predict which ADR technology domains will grow and develop best; there are already a number of companies initiating various projects for removal of space debris and exploiting those market opportunities. In a long-term future, parallel with the deployment of mega-constellations, there may emerge a number of companies providing such services.

The on-going dialogues among industry, scientists, policy makers and economists should give more information on: "how extreme can future events be?", "what is the expected frequency of such events?", and "what damage can be expected?". Realistic and adequate answers to these questions are the silver bullets for defining an appropriate resilience enhancement strategy including governance options for the New Space sector. It should be noted that any steps taken to reduce long-term risks also minimize the potential increase of short-term risks.

These arguments lead us to the fact that the space debris peril has in practice created a number of new opportunities not only in technology development aspects and the creation of a niche market notably in Research & Development and Insurance sectors, but also it is acting as a platform for international dialogues aiming at establishing new legislation and policies for a sustainable global future in the New Space era.

Index